AF713574

Nuclear Tsunami

Nuclear Tsunami

The Japanese Government and America's Role in the Fukushima Disaster

Richard Krooth,
Morris Edelson,
and Hiroshi Fukurai

For my Pal Paul
~ Richard
5-9-15

LEXINGTON BOOKS
Lanham • Boulder • New York • London

Published by Lexington Books
An imprint of The Rowman & Littlefield Publishing Group, Inc.
4501 Forbes Boulevard, Suite 200, Lanham, Maryland 20706
www.rowman.com

Unit A, Whitacre Mews, 26-34 Stannary Street, London SE11 4AB

Copyright © 2015 by Lexington Books

All rights reserved. No part of this book may be reproduced in any form or by any electronic or mechanical means, including information storage and retrieval systems, without written permission from the publisher, except by a reviewer who may quote passages in a review.

British Library Cataloguing in Publication Information Available

Library of Congress Cataloging-in-Publication Data

Krooth, Richard.
Nuclear tsunami : the Japanese government and America's role in the Fukushima disaster / Richard Krooth, Morris Edelson, and Hiroshi Fukurai.
pages cm.
Includes bibliographical references and index.
ISBN 978-0-7391-9569-7 (cloth : alk. paper) — ISBN 978-0-7391-9570-3 (electronic) 1. Fukushima Nuclear Disaster, Japan, 2011. 2. Nuclear energy—Government policy—Japan. 3. Nuclear power plants—Japan—Safety measures. 4. Nuclear power plants—Japan—Design and construction. I. Edelson, Morris, 1926– II. Fukurai, Hiroshi. III. Title.
TK1365.J3K76 2015
363.17'990952117—dc23
2014041979

∞ ™ The paper used in this publication meets the minimum requirements of American National Standard for Information Sciences Permanence of Paper for Printed Library Materials, ANSI/NISO Z39.48-1992.

Printed in the United States of America

One hopes for a
Japanese Spring

With Renewal and Reinvention,
Rebuilding and Rebellion

Against Familial Oligarchies
And Their Lust for Lethal Nuclear Plants—

As well as Against U.S. and French
Pressures to Retread the Terrible
Schemes of Cold War Years

Contents

Acknowledgments

This study emerged from several years of the authors' cooperative work and discussions with other scholars, legal researchers, and environmental activists.

An interlocked web of scholars and witnesses also have gone before us, shaped our thinking, and given us support. So here we would like to thank those who contributed their expert knowledge in the period before and during the moments this particular manuscript took formation.

Oral historian, professor, and essayist Ann Baxandall Krooth generously provided us with research discoveries. Sumico Ishikawa Van Hoy uncovered front-line stories during the ongoing Fukushima nuclear tsunami. Chikako Fukurai and her children lived in Japan in the Fukushima danger zone off and on during the aftermath of the disintegration and meltdown of the Daiichi nuclear plant. Dozens of reporters risked their safety and very lives attempting to accurately report the ongoing lethal events reported in this book. And those who now suffer from irradiation, thyroid cancers, and worse report what then and now has taken place.

The backdrop for this Daiichi disaster is no easy story to tell, as it links World War II to Fukushima today.

The dangers of radiation have long since been known, since the days of Madam Curie. The horrors were seen in 90,000 feet of raw film recorded during the aftermath of the U.S. atomic bombings of Hiroshima and Nagasaki, today preserved at the National Archives located in College Park, Maryland (labeled Number 342 USAF).

Albert Einstein, nuclear experts Dr. John Goffman, Dr. Ernest J. Sternglass, and dozens of other biologists working in the U.S. National Academy of Sciences have exposed the dangers of low-level radiation. John J. Berger's study of *Nuclear Power* (1976) also documented the lethal environmental consequences of the release of nuclides by burning plutonium to boil water to drive turbines to produce electricity.

Others have stressed the shortsightedness of war-making with nuclear and other bombs causing mass destruction, just as effective as subtle irradiation of millions from nuclear power plants.

Historian William Craig tells in *The Fall of Japan* (1967) of wrenching human suffering following the U.S. nuclear bombings of Nagasaki and the fire-bombing of Tokyo. Kurt Vonnegut Jr. recounts in *Slaughterhouse Five* (1969) the burning of life-giving oxygen snuffingi out the life of near one million souls during the U.S. deadly bombing of Dresden.

British air commanders still boast of deploying the American-Wing to bomb German working-class districts to bring Hitler's *luftwaffe* to the peace table. And the great German sociologist Hans H. Gerth recounts the ways in which both Deutschland and Nippon, with ambitions to become master races, were blown to pieces.

To say we are grateful to those seeking to end all forms of nuclear power would be an understatement. Our very lives depend on learning from their experiences, their stories, their insights and warnings. To ignore them will be to our peril.

Here we also acknowledge the help of many contacts and sources, and wish to note some of those whose work or words were extremely helpful

This study of the Fukushima nuclear disaster set off by the Earth's crust movement not far offshore of northeast Japan differs from two admirable volumes taking on the same and related topics. The Union of Concerned Scientists published this past year (2014) something of a science fiction thriller quickened by attaching the many mishaps at the Daiichi nuclear plant to dates and times. This fast-paced encyclopedia mixes horror, suspense, and much explanation of the machinery and metrics of the nuclear world.

A second study from the National Academy of Sciences was still in pre-publication status in late 2014. The aim of the distinguished panels of scientists apparently is to analyze operations and safety issues of nuclear energy plants and to recommend best approaches for environmental—and human—protection.

The two studies, considered with all due respect, may exemplify problems with the discussions about and use of nuclear fuels, problems which could be discussed in two parts, one having to do with energy, the other with the application of nuclear power, broadly considered. As such topics come into focus, questions of motivation and money arise, as will be seen in our study. In short, we are not merely "concerned," but also disenchanted about the corporate triumphs within the Japanese nuclear industry. We are members of an emerging international academy suspicious of pseudo-science, such as claims that allowing for a little danger, our societies can enjoy clean energy too cheap to meter, all the uses for which will be peace and progress.

Chapter by chapter, we also are indebted to the following authors:

Chapter 2: Trial by Sea, Trial by Fire: Professor Yoko Hasegawa in "About Hiroaki Koide, a Nuclear Reactor Specialist" (University of California, Berkeley, May 7, 2011) sketches the increasing power of nuclear-related corporations and interests and the silencing of critics. Supporting the accuracy of this observation and that nuclear problems are man-made is Aileen Mioko Smith, Director, "Green Action" (Kyoto, Japan, August 2010). Green Action and Reuters confirm that Japanese financial institutions, including the country's top three banks, considered providing more than 1 trillion yen ($12.3 billion) in emergency loans to Tokyo Electric Power, Fukushima operator shortly after the disaster.

Chapter 3: Ingenious Invisible Invaders: The effect of radiation, immediate and long-term: succinct explanation of some scientific complexities are available from Wed sites such as "Health Effect of Ionizing Radiation," available at http://www.cardnm.org/healtheffects. An example is the description of action of radioactive decay emissions, Alpha particles, Beta particles, X-rays, Gamma rays, and Neutrons. These rays and particles are called ionizing because as they tear through living tissue, they knock electrons out of their shells. The resulting changed atoms —called ions — are incomplete and unstable. Their structure and electrical balance have been disrupted. All life on Earth and in the oceans lives with exposure to natural levels of ionizing radiation—high-frequency radiation with enough energy to change DNA. Most such genetic damage heals, but the addition of human-made radiation can make it harder for the body to repair broken genes. Christine Dell'Amore, *National Geographic News*, April 1, 2011.

Chapter 4: Damning the Torpedos at Full Throttle: Water cannot be allowed to touch the thing that is deep inside the [Chernobyl] reactor: about 200 tons of melted nuclear fuel and debris, which burned through the floor and hardened, in one spot, into the shape of an elephant's foot. This mass remains so highly radioactive that scientists cannot approach it. But years ago, when they managed to place measurement instruments nearby, they got readings of 10,000 rem per hour, which is 2,000 times the yearly limit recommended for workers in the nuclear industry. (Ellen Berry, "Lessons From Chernobyl for Japan," *New York Times*, March 10, 2011.)

Chapter 5: Facts About Nuclear Reactors Your Mother Never Told You: "To better understand the process of nuclear energy production and its dangers, recall that isotopes are atoms with more or fewer neutrons. In every nuclear reactor there is both fission of isotopes such as uranium-235, and the formation of new, heavier isotopes due to neutron capture, primarily by U-238.

Most of the fuel mass in a reactor is U-238. This can become plutonium-239 and by successive neutron capture Pu-240, Pu-241 and Pu-242, as well as other transuranic isotopes (Pu-239 and Pu-241 that are fissile, like U-235 and Pu-238 that are formed similarly from U-235.)" The increased energy in isotopes heightens dangers and effects of any mishap or leakage. (Richard Krooth, Interview with Dr. John Goffman at the Law Offices of Melvin Belli, San Francisco, 1967.)

Chapter 8: From Sea to Shining Sea: Radiation without End: In early April 2011, media outside Japan reported that Tokyo Electric was bargaining with those affected by the disaster. Its aim was clear: to build new reactors at the Fukushima complex and to mollify spokesmen for evacuees: "Tokyo Electric announced in early April that four reactors at the center of the crisis at Fukushima Daiichi will never go back into service and included a proposed timetable for constructing and commissioning two reactors—No. 7 and No. 8—at the same complex." (*Washington Post*, April 3, 2011). Katsuya Endo, the mayor of Tomioka town, was one of eight Fukushima prefecture mayors who went to Kan to demand compensation and support for employment, housing and education for the tens of thousands of crisis evacuees. (Reuters, *News*, April 5, 2011). BBC News (April 12, 2011) said that Tokyo Electric faced $23 billion in compensation claims and had lost 75% of stock share value. It also described Tokyo Electric's effort to escape liability by having the disaster classified as a "natural disaster of an exceptional character" ("JP Morgan: TEPCO may face claims of $23.6 billion"" BBC News, April 12, 2011)

Chapter 9: Nature as Nuclear Trash Bin: The Los Angeles Times made the point that although protests were directed against Tokyo Electric, design errors it made were less to blame than cozy ties between government regulators and industry executives and complacency about safety. After-disaster human error, confusion, and miscommunication arose from bureaucracies that grew up in the pro-nuclearists' many agencies. (John M. Glionna and Kenji Hall, *Los Angeles Times*, April 29, 2011.) The confusion allowed continuous radiation leaks, a de facto boycott of goods from northeast Japan, government subsidies to this or that repair effort or unproven remedy, and higher utility bills and taxes for the public.

Chapter 11: Returning—On Shaken Feet: Martin Fackler, in the *New York Times*, pointed out that tsunamis could be expected, since those of the past were still silently memorialized by stones marking the limit of low-lying danger zones. "Some places heeded these lessons of the past, but many didn't" (Martin Fackler, "Tsunami Warnings, Written in Stone," April 20, 2011). The slow pace of cleanup and control efforts angered many, further affronted by the government's continual "keep calm" messages and evidences of previous misplaced trust: In Futaba, a town next to the Daiichi plant, several signs stretching across the empty streets extolled the virtues of atomic energy. "Nuclear power is energy for a brighter future," read one. Another said, "The correct understanding of nuclear power leads to a better life." (Andrew Pollack, "Japanese Revisit Nuclear Zone While They Can," *New York Times*, April 22, 2011.)

Chapter 12: Popular Rights vs. Corporate Power: The Tokyo Electric Power Company had laid out an ambitious plan which included use of robots and drones and supposedly reduced dangers to cleanup crews. Ben Belson and Steven Lee Meyers described the robots in reassuring detail: "The remote-controlled PackBots used, built by the iRobot company of Bedford, Mass., are 60-pound contraptions with steel arms and tractor treads. They can lift about 30 pounds, go up and down steps, send images back to an operator and carry a hazardous materials kit that senses radiation." Ben Belson and Steven Lee Myers, *New York Times*, Tokyo, April 17, 2011

Some 4,000 of these U.S.-made robots were sold to Tokyo Electric; few were ever used, nor did Tokyo Electric implement other aspects of its projected effort. Even Tokyo Electric managers complained about the actual—as opposed to announced—refitting and repair, the general manager of Tokyo Electric, two months after the meltdown and explosions, complained of radioactive fuel being uncovered, open to the atmosphere, and possibly even lying on the bottom of the reactor, from which point it could penetrate the concrete and leak into nearby land and ocean (Junichi Masumoto, General Manager, memo of May 5, 2011).

Chapter 14: Making a Nation Safer for Profits: Prime Minister Kan seemed to be on an election and public relations campaign, as the reactors leaked and cleanups lagged. His assurances, however, were scarcely credible to the public, but enough to alienate his corporate colleagues and controllers. His utteramces hurt the support of his political party, the Democratic Party of Japan, and prepared the way for his early loss of office to Shinzo Abe, still in office. The media and public gave him lukewarm support at first, since so many prime ministers had come and gone. *The Wall Street Journal* (May 7, 2011) reported that Kan had "requested" the closure of one nuclear complex and a gradual withdrawal from the use of nuclear power. Martin Fackler and Hiroko Tabuch cast doubt on whether Kan could control the nuclear power interests, and public support for him fell as he increased the sales tax, then cut the increase (Martin Fackler and Hiroko Tabuchi, the *New York Times*, May 7, 2014). Kan was out in two years, a record length in recent times, now surpassed by Shinzo Abe.

Chapter 15: Futures Unknown: The study of withheld information by Noreena Herz offers explanation for the fade-out of protest against nuclear power. Officials simply kept information secret. Tensions with North Korea and Chinese claims and nuclear development and profits to be made by atomic technology and fuels in, especially, India and Thailand probably strengthened the hand of nuclearists. (Noreena Hertz, *Eyes Wide Open*. New York: HarperCollinsPublishers, 2013).

The Economist ("Disaster in Japan," April 21, 2011) suggests that nuclear acquiescence is part of the pattern in Japanese culture of deference to those in charge: "Too often, the ingrained deference of the Japanese permits oligarchs to blunder fearless of protest and social strife." *Democracy Now* (June 7, 2011) also reported on delayed release of often-contradictory or unreliable reports, the only thing able to be authenticated in them was that mistakes were made: "Japanese authorities are now admitting the nuclear crisis at the Fukushima Daiichi nuclear power facility was far more severe than they had previously admitted. On Monday, Japan's Nuclear and Industrial Safety Agency more than doubled its estimate for the amount of radiation that leaked from the plant . . ." in the first week of the disaster in March. David Lochbaum, one of the authors of *Fukushima: The Story of a Nuclear Disaster*, published by the Union of Concerned Scientists, said tardy government mini-revelations were like the Japanese government admitting the possibility of gravity (Comment on Japanese Government Report and Disclosure, June 7, 2011, Chico Harlan and Akiko Yamamoto, *Washington Post*, Tokyo).

The seepage of dubious corrections to previous falsehoods is also a means of muting resistance, since it creates and maintains a fogged-up landing for realizations and consensus of scattered groups of critics. The ideological guerrilla warfare has been compared to the back-and-forth about global warming in the U.S., which includes some outright denials of the continuing reactor leaks or rejections of the findings that the reactors even in cleanup are climate impactors, contributing to global warming and extreme weather (Megumi Tanifuji, Nipponnzan Myohoji, Grafton Peace Pagoda).

A special thank-you to Martin Fackler, Tokyo Bureau Chief for the *New York Times*, perhaps the clearest on and closest to the continuing Fukushima story. He has described the unreliable bureaucrat bulletins, the government-corporate coziness, and problems with cleanup and operation plans. Fackler was one of few who saw through the government promise to cut Fukushima radiation in half by 2014: short-lived isotopes such as that of iodine will naturally become inert by that time, although other isotopes, such as cesium-137, remain active for decades. Another *Times* report notes that the Chernobyl reactor meltdown site is only now—after 30 years—being covered over with a 32,000-ton cap. Martin Fackler, "Japanese Split on Fukushima Radiation Cleanup," the *New York Times*, December 7, 2011; Henry Fountain,

"Putting a 32,000-Ton Cap on Chernobyl," *New York Times*, April 28, 2014). Fackler has also described the interlock between Japanese corporations, which Abe says he will challenge (but hasn't) and some of the pro-corporate steps, such as lowered taxes, easier outsourcing, and some diminishment on job tenure. Abe's nationalistic, right-wing politics would indicate that his talk of nuclear reduction is only that, as we hope our book also makes clear.

Credits for technical help are due Marilyn Chapin, Ann McCardy, Tina Mikfarjam, Aldinette Lockett, and other office staff of the Sociology Department at the University of California, Santa Cruz; and, of course, Kathryn Tafelski and other technical editors of Lexington Books.

With the aid offered by these colleagues and friends, we pursued our task of presenting our analysis of the Fukushima nuclear disaster, its origin, and future ramifications. Needless to say, we alone are accountable for the contents, the facts presented, and the conclusions in this book.

Richard Krooth
Morris Edelson
Hiroshi Fukurai
September 1, 2014

Introduction

As the only country on Earth to undergo two nuclear holocausts, at Hiroshima and Nagasaki near the end of World War II, Japan quickly and decisively denounced nuclear warfare. Yet less than ten years after the war ended, Japan embraced national atomic policies and started a nuclear energy program.

This change in perceptions and programs took place in the context of emergent U.S. post–war global strategies—the essential core of this account.

The "Atoms for Peace" program that emerged during the early stage of the Cold War recognized Japan as a key ally, legitimizing and promoting the peaceful use of nuclear energy. President Eisenhower's Atoms for Peace agenda supplied nuclear-related information and equipment to hospitals and research institutions within the U.S. and throughout the world.

Japan was viewed by the U.S. as the most ideal, newly resurrected state to initiate the building of a nuclear strategic alliance in collaboration with Western European countries. This alliance was designed to fight against actual, potential, and aggressive expansion of the Soviet Union and its socialist allies in Central and Eastern Europe, as well as in China and India during the emergent Cold War.

To lessen or even remove the image of human nuclear devastation from Japanese memories, a massive propaganda campaign undertook to implant the idea that atomic energy could be used for peaceful purposes that would renew Japan's economic and social vitality.

To publicize Japan's national opportunity for the peaceful utilization of atomic nuclear energy, the United States then became the driving force. And toward this end in the early 1950s, the U.S. government made a strategic decision to begin clandestine intelligence operations and psychological programs throughout Japan.

INDOCTRINATION UNPRECEDENTED

Recently declassified documents reveal that CIA and other intelligence operatives played an important role in changing the tide of Japanese public opinion toward nuclear-related matters in Japan.

During the U.S. occupation of Japan under the Supreme Command Allied Forces (SCAP), Japanese political and economic elites were identified. Key American operations made use of Japan's centralized corporate media to appease the world outlook of the nuclear-fearful Japanese citizenry and offset views of anti-U.S. labor forces, grassroots organizations, and the newly installed governmental authorities. Hidden and open persuasion worked; the mission was accomplished by 1954 when the Japanese government budgeted 230 million yen for the nuclear energy program. In 1955 Japan began a national nuclear energy program and built a nuclear plant in Tokai-mura, north of Tokyo, in 1956. By 1957, the plant was producing electricity from nuclear fusion. In the following years, other reactors came online, and the nuclear mechanics developed hotter, more powerful motors inside the original, rapidly aging chassis.

By 1960, General Electric had introduced the blueprint for Light Water Reactors, using normal water supplies, for Japanese nuclear energy plants. These are the cheapest and easiest reactors to build, so heavy industrial firms could help construct each nuclear facility. These also have complicated pipe connections, as salt or alkaline water tends to corrode them.

At the site for the construction of the Fukushima Daiichi ("Number One") Nuclear Energy Plant, GE retained intellectual property rights to the nuclear power unit supplied to reactor Units 1, 2, and 6, while Toshiba supplied structural materials for reactor Units 3 and 5, and Hitachi for Unit 4. The six followed the GE design and were built by Kashima, another multinational Japanese corporation. The lines of authority could be as complex as the piping.

Age of the Daiichi plant is central. Unit 1 was built in 1968, commissioned in 1971, and has been operating for more than 40 years. It was supposed to be decommissioned January 2010; yet, the Japan Nuclear and Industrial Safety Agency (NISA) approved its operation for ten more years by the Tokyo Electric Power Company (TEPCO). This was despite the fact that on February 26, two weeks before the earthquake and tsunami, many organizations, including the Fukushima Environmental Protection Organization (Fukushima Shizen Hogo Kyokai) and the Green Fukushima of Future (Midori no Mirai Fukushima), submitted to TEPCO a petition to disclose closely held information on safety checks and inspections of the Fukushima plant.

Rumored and reported dangers, questioned as early as 1972, were clear. Even though new nuclear powers came online, TEPCO continued to rely on the ancient plant to generate electricity. In effect it was speeding without seat belts or airbags in a vehicle designed before safety equipment was required, and—to extend the analogy—distracted by on-board computers and cell phones chattering about markets, production goals, supplies, and profits.

Japanese officials passed the buck: public revulsion and U.S. pressure forbid nuclear power research, so Japan itself does not produce or design reactors. Only in 2007 did a GE-Hitachi joint venture company emerge in the U.S. to co-develop a new generation of nuclear reactors.

But even the corporation-friendly U.S. Nuclear Regulatory Commission (NRC) said in 1972 that the Fukushima design flaws meant it never should have been licensed. In 1985, the NRC went up an octave: there was a 90 percent chance that a severe accident would collapse the spent fuel-rod containment. Nothing was done. The tocsin sounded in 1976 by three GE nuclear scientists turned out to be their funeral knell. They pointed out that the practice in Fukushima, sanctioned by GE, of putting the spent fuel rods pool on an upper floor, not ground level, of the reactor building, would make it all but impossible to fill the building with cooling water to stop the rods from melting. They were forced to resign.

On March 16, 2011, one of the three scientists, with tears in his eyes, said on CNN that the condition of the Fukushima plant had worsened beyond his earlier dire warnings. The top-floor storage fulminated after plant cooling systems went down and containment cracked. In the crisis TEPCO sent in helicopters to scoop up and dump seawater on the plant, hoping vainly that water would get to the used-rod evaporating pools in the six Fukushima reactors. This drama played well in some media but its epilogue contained more threats: the fuel rods reached melting temperatures, steam from plant coolant and the helicopter drops spewed radioactive steam, and some radiated seawater washed back into the ocean. Meanwhile, GE, from its own sea of half-truths and falsehoods, splashed over the situation claims that the design was sound, evidenced by its use in the construction of 23 of 104 nuclear reactors operating in the United States.

In the early days of the ongoing disaster, one would hear "this is no Chernobyl." True, the names have been changed, but Japan and its people face the future of irradiation; burial of the Daiichi nuclear plant; likely permanent displacement from much of northeast Japan; unrelenting fear and anomie of a displaced population of thousands of people, many with radiation poisoning; and a trophic cascade of irradiated plant, animal, and aquatic life. China is not far; winds blow toward it from the Pacific; and, in general, the environment of Earth is taking another hit.

THE FRENCH CONNECTION

The French link to Japanese nuclear power plants is unique and dangerous for the French, their environment, and their economy.

The French nuclear fuel company Areva produces the ultra-powerful uranium-plutonium mixed-oxide fuel (MOX), used at the Unit 3 reactor in Fukushima

TEPCO and other Japanese power companies commissioned Areva to process MOX fuels, under fire since their initiation by nuclear experts because of a low melting temperature that potentiates lethality. If cooling fails, the MOX melt will release plutonium that threatens public health by raising the risk of fatal carcinogenesis and other cell mutation.

The rush to the crippled Fukushima plant by five French nuclear experts was followed on March 31 by President Sarkozy and Areva CEO Anne Lauvergeon with other than humanitarian motives. The media played these expeditions up and made much of the French advice about disposing of the contaminated water leaking into the ground and the sea. Less noted was that some 80 percent of all French electricity is generated from nuclear plants powered by the plutonium-uranium mix. Any significant disaster might also threaten the French government's nuclear policy in the future.

Fukushima may have even wider economic impact. The government of India, negotiating with GE for the construction of nearly $150 billion worth of nuclear plants, is now reconsidering the agreement and redrafting its own nuclear plans. The U.S. government promoted GE's nuclear investment and exports as part of a foreign policy to export U.S. proprietary technology, especially to the Third World.

PROPRIETARY TECHNOLOGY; LEGAL RESPONSIBILITY

The contract between TEPCO and GE places the liability for any disasters squarely on TEPCO as the operator of the Fukushima nuclear plant. As is common in nuclear operations, TEPCO also carries insurance with over one billion dollars in coverage and guarantees that the national government will pay the remainder of any future legal liability. The exception from this binding responsibility stipulates that the original design of the nuclear plant must be proven to be flawed. The liability of the disaster then shifts to the designer of the nuclear plant, i.e., to GE. If flawed design is proven, GE may be liable for payment of billions of dollars to millions of disaster victims in Japan and elsewhere.

And regardless of the legal strategies or maneuvers, it is still possible that TEPCO and the Japanese government might bring lawsuits against GE, which drafted the original design of the nuclear power plant in Fukushima.

The U.S. responded almost as quickly as the French. The American government persistently offered logistical help, initially rejected by TEPCO. Many people believe that offers of U.S. help including military personnel stationed in Japan arose solely from concern about the health and social welfare of the Japanese. Some, however, suspected as an additional motive the potential liability of GE and the nuclear policy promoted by the U.S. government. Japan eventually agreed to "Operation Tomodachi (Operation Friends)," which mobilized fourteen U.S. Navy ships, more than one hundred aircraft, and nearly twenty thousand U.S. service members stationed in Japan to support disaster victims in remote areas and islands. Military nuclear experts have also been invited to join the emergency committee to deal with the breach of fuel containments and perhaps the nuclear core reactors at Fukushima.

TEPCO, GE, Avera, and both the Japanese and American governments voiced concern over the extent of potential nuclear contaminations in Japan and other countries, and their possible legal liability, loss of corporate profits and revenue, and compensation to millions of disaster victims.

FUTURES REMAIN UNKNOWN

After Fukushima Daiichi has been technically pacified and demobilized, buried under soil, lead, boric acid and reinforced cement, dangers will remain unknown despite the claims of corporations competing for lucrative sanitization and burial contracts. Graveyarding may take up to 30 years and cost billions, since it will require tear-down, recovery of the fuel rods, the burial, and security for a no-man's-land that stretches 60 miles around the forbidden zone. Bidders for the nuclear neutralization will have no responsibilities for leaks or leaches or airborne dangers occurring before the first frantic days of the failures in function at Fukushima.

On April 28th, a brave global team led by Hitachi estimated that it would take at least three decades to return the site to what its engineers refer to as a "green field" state. Competitor Toshiba, Japan's biggest supplier of nuclear reactors, retorted it could take as little as 10 years.

Hitachi spokesman Yuichi Izumisawa said its engineers expected removing the nuclear fuel rods from the plant and packing and transporting them to a safe storage facility would take a decade. Only then could the dismantling of the plant's structures begin, followed by the mop-up of the remaining radiation. "You basically must dismantle the plant from the inside, which is still very radioactive," Mr. Izumisawa reported. "At Hitachi, we are baffled over what kind of technology would allow everything to be finished in 10 years."

In April Toshiba and Hitachi submitted dismantlement schematics to Tokyo Electric Power Company. As no nuclear reactor had ever been fully decommissioned, let alone the four at Fukushima Dai-ichi, the fate of Japan's entire northeast coastline of Tohoku was in question.

Hitachi, the country's second biggest supplier of reactors, had a ready team of 50 experts working on its dismantling plan. It held a joint nuclear venture with General Electric and was also working with the American nuclear operator Exelon and Bechtel, a giant engineering company. Toshiba's team included engineers from Westinghouse, whose majority owner is Toshiba, and the Babcock & Wilcox Company, an energy technology and services company that handles hazardous materials disposal. The two companies had helped shut down the damaged reactor at Three Mile Island.

Comparisons are not only camp, but when it comes to nuclear reactor cleanup they are near meaningless. Still, the cleanup after the 1986 Pennsylvania's Three Mile Island blowout took 14 years, cost some $1 billion, and involved just one reactor during a partial meltdown of the nuclear fuel rods, though the chamber holding them did not rupture. That left two surviving reactors set to be torn down in 2014, with nowhere on Earth to safely store their used plutonium rods. Ukraine's Chernobyl is dead and forgotten despite the explosions, fire and massive radioactive plumes suffusing the atmosphere. Though its reactor is a sundered skeleton covered with sand and lead and entombed in concrete, it will forever release radiation in an area totally uninhabitable.

Perhaps efforts in Fukuyama have created ripples of hope destined to be dashed by stony realities. Tetsuo Matsumoto, professor of nuclear engineering at Tokyo City University, said that the decommissioning process will take more or less time depending on the state of the nuclear fuel: "It could take ten years, or thirty. Will the fuel retain its rod form, or [as in Chernobyl] have melted and collapsed into a big mass? No one will know until the reactor is opened." Only a future generation will have all the facts and the determination or not to fashion a stronger leash on nuclear proliferation or even halt the risky operation of today's reactors.

I

Roots—U.S. Psychological Warfare Strategies and Propaganda Campaigns in Japan

We need to set the framekwork for the fact that within ten years after the detonation of two atomic bombs in Hiroshima and Nagasaki, it seems ironic that the Japanese citizenry would suddenly embrace the use and development of nuclear technology.

Rather than staging a fierce antinuclear stance, the Japanese people made a complete turnaround in their public opinions and attitudes toward the use of nuclear power, particularly in their desire for the peaceful utilization of atomic energy. Such a dramatic shift in opinion was no accident, but the result of a series of carefully orchestrated psychological propaganda campaigns initiated by the CIA and other related U.S. intelligence agencies in the early 1950s.

The first chapter of this book examines the sociopolitical history of postwar U.S. intelligence operations and strategic psychological warfare in Japan; it presents a critical analysis of the process through which the U.S. government was able to mount massive psychological propaganda campaigns in postwar Japan, transforming the atomic-wary Japanese public into a mass of pro-nuclear enthusiasts who came to embrace the use of atomic energy and the development of the country's own nuclear program.

The execution of these propaganda and intelligence programs was so successful that it had achieved its desired effect in less than ten years after the 1945 atomic explosions. Furthermore, it has been revealed that the CIA colluded with a small circle of Japan's power elites—including media moguls who had control over a vast portion of Japan's print and electric enterprises—through whom these clandestine operations were made possible. These intelligence projects provided them the opportunity to advance their own personal and political objectives.

First we examines the relationship between the CIA and its key asset, Matsutaro Shoriki, the owner and president of Japan's largest circulation newspaper, *Yomiuri Shimbun*, and Japan's first commercial television station, Nippon Television Network (NTV). Shoriki's role and connection to the American government and its inner circule are also analyzed.

Our next focus is on America's political interferences. We need to understand the extent of the U.S. influence on the Japanese government, we examine its policy-making prior to the deployment of massive strategic intelligence warfare in the 1950s by CIA and other related intelligence operatives and agencies. Our analysis reveals the pervasiveness of continued U.S. influence and the significant impact of operations in Japan to cultivate fertile ground for the American government to advance its economic and financial interests in Japan and other regions in East Asia.

We reveal the deep-rooted relation between the CIA and the Japanese cabinet. This is followed by concluding remarks about the impact of U.S. covert psychological warfare campaigns on the formation of Japanese perceptions and governmental nuclear policies that allowed the construction of the nuclear power plants in a once nuclear-ravaged Japanese island.

Chapter One

Transformation of the Most Nuclear-Fearful Population to an Atomic Enthusiasts

America's Psychological Warfare Strategies[1]

On March 8, 2011, three days before the devastating earthquake hit Japan's Tohoku region, Tokyo Governor and prominent ultra-conservative Japanese politician Shintaro Ishihara urged the Japanese public that the government should consider developing nuclear weapons within a year. They stated: "All our enemies—China, North Korea, and Russia—all close neighbors—have nuclear weapons; . . . diplomatic bargaining power means nuclear weapons. All the [permanent] members of the [United Nations] Security Council have them."[2] These UN Security Council countries included the U.S., Britain, France, and Russia, all of which succeeded in the development of nuclear weapons by 1960, as well as China, which detonated its first successful nuclear bomb in 1964.[3]

Ishihara's call for Japan's nuclear armament is a typical view shared by neo-conservative politicians, that Japan's unconditional surrender to the Allied forces in 1945 demoted Japan's prewar reputable status into the rank of second-class nation-states in the world community, and Japan's nuclear rearmament must be viewed as an indispensable strategy to restore Japan's once-revered world status. Indeed, the postwar Japanese government always maintained a strong desire to rebuild its military with powerful weaponries including nuclear bombs.

The historical root of such imperial political aspirations goes back to the mind-sets of politicians who helped construct the modern sovereign state of Japan in the late nineteenth century. Such a persistent desire to attain and maintain a strong military has been always part of the national interest—since a select group of power elites declared Japan's political sovereignty and helped establish the modern state in 1868. This newly established Japanese state government soon became one of four permanent League Council states of the League of Nations, along with Great Britain, France, and Italy, thereby achieving reputable political status and establishing its prominent political privileges in the world.

After the defeat in World War II, the restoration of its prewar international reputation and equitable political status became a paramount concern of the postwar Japanese government. As the five permanent members of the United Nations Security Council had all successfully built their own nuclear program and developed their nuclear weapons, the development of atomic arsenals and subsequent nuclear armament became an integral part of Japanese governmental agendas and policies. Nonetheless, the Allied forces led by the U.S. initially tried to prevent Japan from becoming a nuclear powerhouse by developing its own nuclear programs.

In order to secretly begin the development of nuclear weapons behind the influence of the U.S. government, Japan attempted to put forth a governmental project to form a secret alliance with the government of West Germany in the late 1960s.[4] During World War II, research projects had been conducted by the military to develop nuclear weapons in Germany and Japan.[5] Japan's scientific collaboration with the West German government was considered crucial because, during the military occupation of Japan by the Supreme Command Allied Powers (SCAP) from 1945 to 1952, the SCAP ordered the complete destruction of Japan's prewar scientific and experimental infrastructures used for the development of nuclear weaponry in both private and public research labs, including experimental cyclotrons used by Japanese scientists to engage in nuclear research and development.[6]

The Allied Forces' destruction of each and every nuclear research lab continued, despite Japanese physicists' assurance that they would use research labs or cyclotrons only for nonmilitary applications.[7] The ban on all research and development of nuclear energy was lifted only after the San Francisco Peace Treaty went into effect in April 1952.[8] By that time, the U.S., the Soviet Union, and Great Britain had already developed their own successful nuclear programs.

In order to jump-start the development of nuclear weapons, former Administrative Vice Foreign Minister Ryohei Murata had drafted a detailed governmental plan to sign a secret pact with West Germany in order to rebuild Japan's own nuclear weapon program.[9] He previously served as an ambassador to Germany and acted as an adviser to the Minister for Foreign Affairs. Based on the government plan, Japan made an initial approach to the West German government to secretly engage in the development of nuclear weapons nearly a quarter-century after the end of World War II.

On June 2, 1969, at a hotel in Hakone, a location of a noted Shinto shrine called the Hakone Gongen in Kanagawa Prefecture, high-ranking officials of both West German and Japanese governments met in order to work on a collaborative project to build atomic bombs. Before the meeting in Hakone, Murata visited the West Germany Embassy in February 1969 and made the arrangements necessary to hold the secret meeting in Japan. Three representatives from the Japanese government included Ryohei Murata, as well as international research director Takashi Suzuki and subsection chief of analysis Hisahiko Okazaki. The German government was represented by policy project director Egon Bahr and two other high-ranking officers.[10]

The Japanese side had expressed to German officials the necessity to build the nuclear weapon program in order to attain greater sovereignty and political independence, suggesting that the secret collaborative project to develop nuclear weapons was imperative to regain superpower status in the world community. The Japanese officials also explained that Japan had already acquired necessary proprietary knowledge in the production and enrichment of plutonium. The fuel was generated as fission products by nuclear reactors at electric power plants on designs imported from the U.S. government. The missile delivery technologies had already been developed by Japan's aerospace program. The German representatives agreed with the importance of seeking greater political independence from the U.S., but finally decided not to participate in the project, as they cited that political situations in West Germany were radically different from those in Japan, as their homeland was politically severed into two sovereign East and West territories, each of which was placed under the direct control of the Soviet Union and the U.S., respectively.[11] The bifurcated political control over Germany by the two superpowers made it unlikely that they could create the political conditions or supply resources necessary to carry out such projects.

It is ironic that the Japanese government had always denied the existence of such a secret project to build nuclear weapons. Prime Minister Eisaku Sato, for example, announced in 1968 Japan's Three Nonnuclear Principles (*Hikaku San-Gensoku*), declaring that Japan was not to make or possess nuclear weapons, and not to allow their entry into Japan. Since the end of World War II, Japan's quest to build nuclear weapons always had been hidden by the political drives of its leaders, and the government continued to develop its own nuclear project—first initiated to promote the peaceful utilization of atomic energy in early 1950s.

It is widely acknowledged that the U.S. government had already recognized Japan's desire to acquire technical know-how necessary to produce nuclear weapons. The U.S. also regarded Japan as an important political ally in the formation of the united front against the Soviet Union in the midst of Cold War, as well as a future consumer of America's proprietary nuclear technologies such as nuclear reactors, nuclear-fueled electric power plants, medical technology, food processing, and agricultural applications. At the same time, the U.S. government had recognized the presence of strong anti-nuclear political movements by Japan's civic activists, teachers' unions, and others grassroots organizations against the development, use, and deployment of atomic weapons in Japan.

In order to create fertile ground for the exportation of nuclear technologies into the Japanese market, the U.S. government then decided to unleash an unprecedented level of intelligence operations and psychological strategic propaganda campaigns and programs in Japan to transform public opinion from strong anti-atomic sentiments to pro-atomic attitudes and opinions. The Central Intelligence Agency (CIA) and other U.S. intelligence agencies in Japan cultivated their Japanese assets and solicited their support to their propaganda campaigns. *Yomiuri Shimbun* owner Matsutaro Shoriki, who later became the president of Japan's first private television station, played a prominent role in the U.S. psychological warfare campaigns. Japan's mass media then served as an effective system for building consent on the use of nuclear energy. The result was the nationwide adoption of the belief, called the "safety myth" (*Anzen Shinwa*)—that the use of nuclear power and energy in Japan was absolutely safe. A complete turnaround of public opinion was achieved in less than ten years after nuclear bombs had exploded in Hiroshima and Nagasaki in 1945.

MATSUTARO SHORIKI AS CIA "ASSET" CALLED THE PO-JACKPOT

One of the CIA's key Japanese allies was Matsutaro Shoriki, who had engaged in prewar propaganda campaigns in order to advance his own economic ambitions and political interests.[12] The CIA considered him as one of the most valuable "assets" for the execution of the CIA psychological warfare operations.[13] How highly the CIA regarded Shoriki in the intelligence operation was evidenced by the CIA's secret code name given to him—PO-JACKPOT—indicative of his integral role in the successful execution of propaganda operations in Japan.[14]

With the onset of the Cold War and increased threats of communist influence in the world, the U.S. government had relied on multiple sources, including former military intelligence officers, propaganda operatives, and even university faculty members, in order to analyze intelligence information and draft psychological warfare programs. In the early 1950s, Undersecretary of State James Webb asked then MIT President James Killan to assemble a team of scholarly scientists to make critical contributions to academic research on U.S. foreign policy.[15] The CIA then funded the establishment of an international research center at MIT in 1952 to work on clandestine psychological warfare programs under the code name of "Project Troy," aptly named after Homer's Greek epic.[16] The project relied on American scientists and university faculty from both MIT and Harvard to help draft U.S. policies on international communication, formulate economic development and political programs, and critically analyze the communist movement and anti-West activities in the world.[17]

The "Atoms for Peace" program was devised to ensure that the postwar West, including Japan, would accept strategic plans created by the U.S. government and American corporate sector that promoted the civilian uses of nuclear fuel and technology, including nuclear power.[18] The program was also designed to subordinate these states to U.S. nuclear policies and new technologies, making them potential corporate clients and consumers of uranium fuels and proprietary technologies developed by U.S. nuclear industries and corporations.

Given that Japan had been the only victim of devastation of atom bombs ever used against civilians, the U.S. government had recognized that Japan's acquiescence to America's nuclear policies and programs was symbolically crucial for the successful promotion of the "peaceful" use of nuclear energy in the world. Henceforth, Japan was targeted by U.S. intelligence for the dissemination of propaganda and psychological campaigns in the early 1950s.

Japanese politicians also welcomed such opportunities to develop Japan's nuclear program and eventually their own nuclear arsenal. Yasuhiro Nakasone, who later became the Japanese prime minister in the early 1980s, visited the U.S. in 1953 to examine the use of nuclear energy and its commercialization; and successfully persuaded the Japanese cabinet in 1954 to create the first budget for research and development of nuclear energy and the construction of nuclear experimental plants.[19] In the following year, he helped to introduce and pass eight bills for nuclear research and development, including the Atomic Energy Basic Law on December 16.[20] For his significant contribution to the development of Japan's nuclear program, Nakasone was awarded the "Atoms for Peace" prize by the World Council of Nuclear Workers (WONUC) and the Japanese NPO International Institute for Sustainable Peace (IISP) in 2008.[21]

In order to transform Japan's strong antinuclear sentiments into pro-atomic attitudes and opinions, the U.S. intelligence agencies turned to one of Japan's most powerful media executives, Matsutaro Shoriki, owner of the largest circulating newspaper in Japan, *Yomiuri Shimbun*. There was also a close interpersonal relation between Nakasone and Shoriki, such that Nakasone stated in his memoir that he had promoted nuclear power and atomic energy programs in close collaboration with Shoriki, ever since participating in a summer seminar at Harvard University led by Henry Kissinger.[22] Shoriki was soon to become the president of the first commercial TV firm called

Nippon Television Network. In short, Shoriki held unequivocal control of Japan's two most successful print and electric media institutions. In order to approach Shoriki, nonetheless, the U.S. first approached his personal secretary and made him the CIA's "asset" in its first step for the execution of psychological warfare operations in Japan.

THE CIA AND MATSUTARO SHORIKI

Shoriki's personal secretary Hidetoshi Shibata became a key CIA informant and provided the CIA crucial access to a circle of powerful Japanese policymakers during the CIA's earliest phase of psychological operations in Japan.

Working previously as a radio commentator, Shibata became the key intermediary between Washington and Matsutaro Shoriki, who later became the director of Japan's domestic nuclear energy program.[23] Shibata compiled massive internal documents, including a list of an exclusive group of privileged elites in Japan's political and economic circles. As a CIA informant, he frequently visited the U.S. and had contacts with key government officials, representing various corporate interests and strategies.[24]

In the prewar era, Shoriki had served as a police commissioner to the Tokyo Metropolitan Police Department in the powerful Interior Ministry that directed the state's brutal repression of domestic communist activities, socialist workers' unions, and labor demonstrators and organizers.[25] In 1924, Shoriki acquired the newspaper *Yomiuri Shimbun* with the aid of Tokyo Mayor Shimpei Goto.[26] At the time of purchase, it had only a 55,296-copy daily circulation, but with extraneous political and economic assistance and support, he was able to rebuild the company, and in 1954, it had a daily circulation of over two million.[27]

Shoriki held various key government positions and cultivated important political networks with the police, prosecutors, and political officials. For example, he played a key role in the anti-democratic Imperial Rule Assistance Association (*Taisei Yokusankai*), held a seat in the Imperial Diet, was an adviser to the wartime Koiso Cabinet, and actively supported Japanese colonization of Manchuria in China.[28] Shoriki was involved in virtually every major incident of police repression up to 1923, including the Waseda University sit-in of 1917, the Tokyo rice riot of 1918, and the 1921 crackdown on the Korean community following the assassination of pro-Japanese Korean leader Min Won-Sik in the Tokyo Station Hotel.[29] His anti-Korean and anti-communist stance became his defining characteristic, as he successfully spread false rumors, right after the Kanto Great Earthquake in 1923, alleging the Korean people's riots and widespread looting justifying the massacre of nearly six thousand Korean residents by the hands of Japanese militias in the regions of Tokyo.[30]

In 1936, he helped efforts in the merger of the news agencies into one governmental propaganda organ, *Domei Tsushinsha*, to be placed under the direct authority of the Ministry of Communication.[31] This centralized government bureau controlled the production of state propaganda and dissemination of censored foreign and domestic news through its extensive network of radio stations in Japan and East Asia.[32] His conservative ideology, state-centered policies, and anti-communist activities prompted the imperial government's ideological warfare and propaganda campaigns against the public. When the imperial government had called for censorship of certain truths and publication of falsified information, Shoriki's newspaper led the way. Shoriki also exercised rigid control over employees to raise productivity while cutting their wages.[33] His newspaper then became the main mouthpiece of the state for the support and maintenance of Japan's military dictatorship in the 1930s and 1940s.[34]

He was also known as an excellent event promoter, as he successfully organized a Japanese tour for Babe Ruth and the New York Yankees in 1935, followed by his successful establishment of the Japanese equivalent of an American-style baseball team, named the Yomiuri Giants, in Tokyo in 1936.[35] He became the first commissioner of the Nippon Professional Baseball (NPB) in 1949, helped realign the Japanese Baseball League into its present two-league structure, and sponsored the Japanese equivalent of America's World Series in baseball.[36]

After Japan's 1945 surrender, Shoriki was soon arrested, classified a Class-A war criminal, and imprisoned in Tokyo's Sugamo Prison for two years.[37] With the emergence of the Cold War with the Soviet Union and perceived threats of communism in East Asia, he was released without trial in 1947, and his virulent anti-communist stance helped to redeem him in the eyes of the U.S. intelligence communities and conservative politicians in Washington.[38] As a media mogul and an ambitious politician, Shoriki proved a natural asset for the CIA's covert propaganda campaign. In addition to his CIA code name of POJACKPOT and PODAM,[39] his *Yomiuri Shimbun Newspaper* was also coded as POBULK, while the CIA branch station in Tokyo was called PODIVA.[40] The CIA also helped establish *Naikaku Joho Chosasitsu* (Cabinet Research Chamber, dubbed as the "Japanese CIA") situated within the Japanese Cabinet in 1952 and gave it the cryptonym of POLUNATE.[41]

One of the most important operations by CIA and other related intelligence and information agencies was the establishment of Shoriki's Nippon Television Network (NTV) in 1952, the year that the Allied Occupation of Japan formerly ended. The CIA and the U.S. Information Services (USIS) regarded Shoriki's construction of Japan's first privatized television network to be crucial for its propaganda campaign programs.[42] Washington also considered the role of the TV media to be useful in promoting pro-Western ideologies, while offsetting any ill effects of Japanese sentiments produced by the U.S. military occupation of Japan.[43]

Recent government documents revealed the deep involvement of the following three groups in the establishment of the NTV, Japan's first commercial television corporation: (1) collaborative Japanese groups, consisting of Matsutaro Shoriki as the president of *Yomiuri Shimbun* and NTV and his private secretary, Hidetoshi Shibata; (2) the American Council on Japan or so called the Japan Lobby, with members including William Castle who was a former ambassador to Japan, and the CIA consultant Eugene Dooman, who was a former counselor of Japan and leader of the Dooman Group, who was also involved in the CIA's psychological warfare programs in postwar Japan, and James Kaufman, an attorney for RCA and General Electric who taught law at the University of Tokyo while serving as an adviser to the CIA at the same time;[44] and (3) the Office of Strategic Services (OSS), a U.S. intelligence agency formed during World War II in order to coordinate espionage activities behind enemy lines, the members of which included James Murphy who was an attorney for Shoriki's NTV and OSS' "number two," and William Donovan who was its Director as well as an adviser to the MSA (Military Security Act),[45] a major vehicle of U.S. foreign assistance to Japan, including military, economic, and technical support.[46]

The OSS served as a predecessor to the CIA; its original mission was to coordinate espionage activities behind enemy lines for the operations of various branches of U.S. armed forces. President Franklin D. Roosevelt established it in 1942 as a wartime intelligence agency to engage in psychological warfare, anti-espionage activities, and violent operations behind enemy lines. Its intelligence branch was then taken over by the newly created CIA that was established by the passage of the National Security Act of 1947.[47] The CIA was created as an operative unit to engage in intelligence activities and was placed directly under the authority of the executive branch of the U.S. government. It had operative missions and programs in Japan and East Asia and their foreign policies and governmental assistance programs were largely shaped and formulated by the Japan Lobby in coordination with Japanese and intelligence groups.[48]

The specific activity of the Japan Lobby involved the creation of pro-Japanese political voices within the U.S. government in order to gain necessary support to former Japanese political and economic elites, who were purged by the Supreme Commander of Allied Powers (SCAP) during its postwar occupation of Japan.[49] Many members of the Japan Lobby were former high-ranking governmental officers who worked closely in foreign services within the Department of State. They engaged in psychological warfare, collaborated with the CIA, and their first effort was directed toward controlling the Japanese media. Shoriki's *Yomiuri Shimbun* and the newly established NTV were to serve as an effective political propaganda machine in Japan.

THE ESTABLISHMENT OF THE NIPPON TELEVISION NETWORK

From its inception, the NTV was not designed to become a popular cultural medium for the dissemination of current events or news information or even to promote commercialism, consumerism, or capitalism. Rather, the CIA's cover-operation was to build an anti-communist military network and engage in propaganda, espionage, and subversion for Japan's postwar planning.[50] Its establishment was formulated by the American microwave engineering firm called UNITEL, which was originally founded by William Halstead, who previously worked for the Office of War Information (OWI), another propaganda organization established during World War II.[51]

As early as 1950, the Japan Lobby and American politicians saw Japan as a potential target of communist influence and suggested that the television network in Japan could serve as an important propaganda weapon in the battle against communism in Asia. And in this effort, right-wing Republican Senator Karl Mundt played an important role in the establishment of the NTV as a crucial propaganda organization in Japan. Previously, Mundt authored the legislation to create the Voice of America (VOA), an anti-communist radio station of the U.S. armed forces in Asia.[52] He sent Henry Holthusen, who served as a consultant to the U.S. Senate Foreign Relations Committee, and also represented the UNITEL Corporation to Japan, in order to meet Shoriki and launch the NTV as a television version of VOA in Japan.[53]

The UNITEL envisioned the creation of the global TV network in promoting the Cold War rhetoric of freedom, democracy, education, science, and culture, predicting that TV would eventually play a paramount role in promoting these American ideologies and values in the face of Soviet propaganda.[54] With the support of the CIA and USIA (the United States Information Agency), the UNTEL was instrumental in the privatization of television in Japan and, in August 1953, helped to establish Japan's first privately owned TV station, NTV Tokyo.[55]

USIA was originally created as the strategic branch office for the United States Information Services (USIS), and its main headquarter was situated inside the U.S. Embassy of Toranomon Bekkan (its annexed building) in Tokyo.[56] The USIS financed, produced, translated, and distributed propaganda films and motion pictures, including *The Yukawa Story*, based on the life of a U.S.-trained nuclear physicist who later received the Nobel Prize. Through TV and movie theaters, the USIS claimed to have reached half the population of Japan though its Motion Picture Branch alone by the mid-1950.[57] USIS-sponsored shows also made up tens of thousands of hours of radio programming, and in 1955 alone, this propaganda agency supported 18,3000 hours of radio programs, or 50 hours per day throughout the year.[58] By 1955, the USIS was also behind the publication of more than one hundred books per year, including many anti-communist bestsellers.[59]

The political use of propaganda was first experimented with and then perfected by the U.S. government in the late 1910s and early 1920s when the Woodrow Wilson administration established a public propaganda machine called the Committee for Public Information, led by a "muckraker" George Creel, which succeeded, within six months, in transforming a pacifist and antiwar U.S. citizenry into a hysterical, war-mongering populace that began to support U.S. participation in World War I.[60] American journalist Walter Lippmann, who first suggested President Wilson establish the propaganda machine, published a book called *Public Opinion* in 1922, which became a revered bible for the public relations industry and military propagandist corps, suggesting that the "manufacture of consent" through the bombardment of propagandas would become effective in the construction of false reality.[61]

Louis Schmitt, number two of the USIA in Japan, later admitted that the U.S. government had identified Japan as the most ideal country to promote a pro-atomic energy campaign. Given that Japan is the only country in the world to suffer from the deployment of nuclear weapons directed against its civilians, if its citizenry was persuaded to use of atomic energy, other nations could be easily persuaded to accept the use of nuclear energy and build their own nuclear programs.[62] In the dramatic words of the U.S. AEC Commissioner Thomas Murray, Japan was deemed as an ideal country for propaganda campaigns because, "now, while the memory of Hiroshima and Nagasaki remain so vivid, construction of such a [nuclear] power plant in a country like Japan would be a dramatic and Christian gesture which could lift all of us far above the recollection of the carnage of those [Japanese] cities."[63]

The TV network project also drew enormous political and economic interest in Japan, as Shoriki raised 700 million yen to help build the NTV and its Japan-wide NTV Network. Three stages of its development were suggested to make the NTV Network an effective outlet for the dissemination of pro-nuclear propaganda, beginning first with the establishment of the Tokyo TV station in its capital city, followed by Nagoya and Osaka TV stations, then finally by nineteen other local stations in various regional locations.[64] UNITEL also sought to link NTV with its television affiliates in Korea, Taiwan, and the Philippines to form the Pacific Network, which would be further connected to a worldwide UNITEL Network.[65]

The UNITEL document suggests that the building of the TV network in Japan was crucial for the U.S. government for four reasons: (1) the creation of a military intelligent network for security forces stationed in Japan; (2) a formation of a strategic defense line against potential communist threats; (3)

campaigns of psychological warfare strategies to engage in dissemination of U.S. propaganda; and (4) an industrial and educational medium to disseminate the American way of life and pro-Western ideologies, thereby making Japan a future consumer of both American culture, manufactured goods, and industrial products and technologies, including nuclear power.[66]

CIA's psychological warfare was designed to transform and eliminate negative Japanese public opinions and attitudes toward atomic-related issues such as nuclear weapons and atomic bomb tragedies. Indeed, the history of the Japanese nuclear program was deeply rooted in the U.S. psychological warfare and strategic propaganda campaigns devised in the1950s; the Pentagon defined the strategic warfare as "the planned use of propaganda and other psychological actions [that] have the primary purpose of influencing the opinions, emotions, attitudes, and behavior of hostile foreign groups in such a way as to support the achievement of national objectives."[67] Similarly, the Department of Defense defined psychological operations as "the planned operation to convey selected information and indicators to foreign audiences to influence their emotions, motives, objective reasoning," with the ultimate purpose to, "induce or reinforce foreign attitudes and behavior favorable to the originator's objectives."[68]

In order to influence the public opinions, emotive perceptions, and general attitudes of Japanese citizens toward nuclear weapons and atomic experiments, the CIA had to engage in near-perfect execution of their psychological strategies in Japan. This was especially important due to the Fukuryu-maru ("Daigo Fukuryu-maru") incident, when, on March 1, 1954, under Operation Castle, the U.S. exploded a hydrogen bomb, Bravo, in Bikini Atoll in the South Pacific. This nuclear blast was supposedly kept as a secret experiment but was exposed to the world by the group of Japanese fishermen on the fishing boat that was operating near the Bikini lagoons.[69] All twenty-three fishermen on board were exposed to a high level of radiation.[70]

The news of Japanese fishermen's exposure to nuclear radiation prompted massive public demonstrations against the continued use of nuclear bombs, further exacerbating anti-U.S. and anti-imperial sentiments all over Japan. Fukuryu-maru Captain Aikichi Kuboyama soon died due to radiation-related diseases on August 23.[71] Nuclear pollution was also detected in fish and vegetables in Japan. The day of the Hiroshima Memorial on August 6th was used to voice anti-U.S. attitudes and criticize the continued use of atomic power by the U.S. government in the world.

On August 12, 1953, the Soviet Union also succeeded in exploding its first hydrogen bomb in the midst of its nuclear arms race with the U.S. Four months later, on December 8, 1953, American president Dwight Eisenhower responded to the Soviet aggression by announcing the Atoms for Peace program at the United Nations. He declared that the U.S. government would release previously withheld atomic-energy-related information to the public throughout the world. The internal document prior to the U.N. Speech revealed that the U.S. had three specific objectives in mind with the release of this information to its allies: (1) evaluating and examining weapons effects and capabilities, (2) assuring the continued use of atomic weapons, and (3) establishing effective response to Soviet atomic capabilities.[72]

The barrage of aggressive pro-nuclear-energy articles and TV programs began with *Yomiuri Shimbun*, NTV, and their affiliated media outlets in 1954. On January 1, 1954, *Yomiuri Shimbun* began a much anticipated daily serial, "Tsuini Taiyo o Toraeta: Genshiryoku wa Hito o Kofukuni Suruka" (We Finally Captured the Sun: Does Nuclear Power Bring Happiness to Mankind?), embracing the peaceful use of nuclear power, suggesting that atomic energy is clean energy and solves the anticipated shortage of energy resources of the future.[73] The purpose of the daily serial was to promote and provide strong support to the Atoms for Peace program initiated by Eisenhower the previous year.[74]

At the same time, the popular cartoon character called Tetsuwan Atom (Astro Boy), who was invented in 1952 by Japan's most popular cartoonist Osamu Tezuka, gained nationwide attention among teenagers. This robotic boy was fueled by nuclear energy, along with his sister Uran (Uranium), older brother Kobaruto (Cobalt), and baby brother Chitan (Titanium). Astro Boy, created by the human scientist Doctor Tenma, always wanted to be accepted by human society. Using his nuclear power, he defeats rogue robots, fights crime, and destroys the evil plots of mad scientists who continued to threaten human society. Ultimately, his good deeds bring peace to human society and lead to people's acceptance of atomic-fueled robots.[75] This particular popular cartoon had appeared in Japan for seventeen years from 1951 to 1968 in a monthly magazine called *Shonen (Young Boy)*, published by Koubunsha in Tokyo.[76] The TV version of Astro Boy was mostly shown by NTV and Fuji-TV affiliates from 1963 to 1966 and again by NTV from 1980 to 1981.[77]

The U.S. intelligence agencies also tried to develop pro-American attitudes and opinions by running stories in *Yomiuri Shimbun* and other media outlets, claiming that U.S. government support led to Hideki Yukawa's Nobel Prize for his prediction of the existence of mesons on the basis of theoretical work on nuclear forces.[78] *Yomiuri Shimbun* also established the Yukawa Scholarship Program in 1950, commemorating Yukawa as Japan's first Nobel Laureate for his work on nuclear forces in fundamental physics. His work on nuclear matters played an important part in the debates about the peaceful implementation of atomic energy in Japan.[79] *Yomiuri Shimbun* also sponsored the special exhibit, "Darenidemo Wakaru Genshiroku (Anyone Can Understand Nuclear Power)" in Shinjuku, Tokyo, from August 12 to 22, 1954, educating the Japanese public about basic knowledge on nuclear power and energy.[80]

Furthermore, in order to transform and eliminate any anti-atomic sentiments still etched deep in the Japanese consciousness, another campaign was unleashed through the newspaper and television network owned by Shoriki to promote a nuclear-related public exhibit in Tokyo based on the "Atoms for Peace" program. Louis Schmitt of the USIA in Tokyo said that the U.S. had to put all its effort to promote this program in order to improve the perception of atomic energy in Japan.[81]

In November, 1955, Shoriki's *Yomiuri Shimbun* co-sponsored the much-hyped U.S. "nuclear exhibit," welcoming the atomic program back to Japan with a Shinto purification ceremony in Tokyo and a speech from Eisenhower, declaring the exhibit "a symbol of our countries' mutual determination that the greater power of the atom shall henceforth be dedicated to the arts of peace."[82] The exhibit also traveled to Hiroshima and six other cities, highlighting and emphasizing the peaceful application of nuclear energy for purposes such as electricity, cancer treatment, food preservation, insect control, and scientific research advancement, while scrupulously avoiding any rhetoric related to its military applications.[83] In Kyoto, 155,000 people attended the exhibit despite snow and rain.[84] The exhibition in Hiroshima attracted more than one hundred thousand people from Hiroshima City and neighboring regions.[85] The exhibit was so successful that officials reported, "the change in opinion on atomic energy from 1954 to 1955 was spectacular . . . atom hysteria was almost eliminated and by the beginning of 1956, Japanese opinion was brought to popular acceptance of the peaceful uses of atomic energy."[86]

Prior to the nuclear exhibit, NTV produced and ran a series of television specials, including *Genshiryoku no Heiwa Riyo* (Peaceful Use of Atomic Energy) on February 2, 1955,[87] followed by the grand showing of the popular American movie, *Genshi Miraisen* [Destination Saturn: Buck Rogers] on March 27, which was originally produced in 1939 by Universal Studios in Hollywood, in which the main character maneuvers an atomic rocket ship to fight aliens.[88] On May 11 and 13, NTV ran a live broadcast of the Public Lecture on the Peaceful Utilization of Atomic Power.[89] On January 2, 1956, NTV ran the New Year Round-Table Discussion, *Genshiryoku o Kataru* (Let's Talk About Nuclear Power),[90] followed by the Public Lecture on Technical Problems of Nuclear Electric Energy on May 17. The lecture was given by Baron Christopher Hinton, a British nuclear engineer and supervisor of the construction of Calder Hall, which became the world's first large-scale commercial nuclear power station in Great Britain.[91]

Perhaps the most representative media propaganda may be found in the editorial of *Tokyo Shimbun*, whose parent company, Chunichi Shimbun, had a close corporate tie with *Yomiuri Shimbun*. On December 31, 1955, its editorial declared that "[with the peaceful use of atomic energy], the cost of electricity will be reduced to 1/2000. With this, neither thermal power generation nor large smokestack or the coal-storage will be needed. In addition, everyday necessity of transporting coals and its wastes back and forth will be superfluous. As long as we can utilize the gas turbine in a closed room, even the water is no longer needed. We do not have to build the power plant in a remote area, but rather, it can be readily established in a basement of a building."[92]

THE NTV, SHORIKI, AND THE WALT DISNEY PRODUCTION

In order to further propagate the nuclear program, a much-anticipated film called *Our Friend the Atom* was created in 1957 by the Walt Disney Company, examining and emphasizing the peaceful benefit of nuclear power and atomic energy. The movie production was financed by both General Dynamics and the U.S. Navy.[93] The story indicates that prominent scientist Dr. Heinz Haber hosted a close look at the "possibility of an exciting new power source. He starts by comparing atomic energy to a genie in a bottle, both of which capable of doing both good and evil, and it is up to human kind to develop safe controls over this largely unexplored science."[94] The film explains in detail the history of discovery of the atom and nuclear energy and demonstrates an atomic chain reaction using mousetraps.[95] It also talks about potential applications of nuclear energy to power electric power plants, ships, submarines, airplanes,[96] and spaceships.

This film was shown to millions of children in public schools, science classes, and movie theaters around the world. Co-founder of Walt Disney Productions and its CEO Roy O. Disney visited Japan in 1957 and asked Shoriki to show the film through his NTV affiliated network.[97] Walt Disney Productions signed an agreement with the NTV in December and the movie was shown to great enthusiasm from the Japanese audience on New Year's Day, 1958.[98] Walt Disney Productions previously promoted the use of nuclear energy in 1954 in another popular film, *Twenty Thousand Leagues Under the Sea*. The story was originally written by French science fiction novelist Jules Verne, in which a huge atomic-powered submarine called *Nautilus* was built in secrecy by Captain Nemo who then traveled the "seven seas," examining and exploring the underwater world around the globe.[99] The world's first nuclear-powered submarine, the United States Navy's USS *Nautilus* (SSN-571), was built by General Dynamics and was named after the fictional vessel, which also appeared in the Disney movie, *Our Friend the Atom*.[100]

In August, seven months after its showing in Japan, NTV also began showing the *Disneyland* which was also produced by Walt Disney Studios and the American Broadcasting Company (ABC).[101] The Walt Disney Studios had produced numerous training, propaganda, and educational films for the U.S. armed forces under the leadership of Robert Carey Broughton, who had previously worked for the Field Photographic Branch of the OSS.[102] During his time with the OSS, he closely worked with Hollywood director John Ford to make a number of documentary war films, including *Battle of Midway*, which won an Academy Award for best documentary in 1942.[103] The OSS extensively utilized "films in intelligence with the OSS Intelligence Photographic Documentation Project. . . . [in order] to establish a worldwide photographic intelligence file of areas of strategic importance."[104] After World War II, he returned to the Walt Disney Studios and worked for Ub Lwerks, who co-created Mickey Mouse and helped produce many movies for Walt Disney Productions, many of which were shown in Japan.[105] During the war, many Disney characters joined the propaganda efforts, as Donald Duck appeared in a number of comical propaganda skits, including the Academy Award–winning "Der Fuehrer's Face" in 1943.[106]

The effects of these pro-nuclear agendas and programs were felt immediately. The USIA carried out a survey in April 1956 to examine the extent of the indoctrination program through the use of exhibits in Japan and found that 56 percent of Japanese approved of President Eisenhower personally. Nonetheless, 60 percent of them still believed that atomic energy would "prove more of a curse than a boon to mankind," while 25 percent believed that the U.S. was "making sincere efforts" at nuclear disarmament.[107]

Sustained CIA programs and USIA activities in the following years, however, were able to produce a significant change in Japanese public opinion. While in 1956, 70 percent of the Japanese public surveyed considered the "atom" as "harmful," by 1958, such sentiment only constituted 30 percent of those surveyed.[108] The massive propaganda campaign was able to persuade the Japanese public that Japan needed to become a modern scientific-industrial power, and that the lack of energy resources in Japan signified a need for the utilization of nuclear power, which was considered to be safe and clean.

The USIA also produced in secrecy—given that it did not want to provoke anti-US sentiments—TV and radio news and commentary shows; it not only funded such news and entertainment programming but also preapproved scripts for them.[109] One of the most influential of the USIA-backed programs was a movie called *Jet Cadets No. 101* (Jettoki Shutsudo Dai 101 Kuko Kichi), released in December 1957, which depicted the training of American-made F86F Sabre jet fighters by Japanese air cadets led by prominent Japanese movie star Ken Takakura. The movie was shown in 2,200 theaters and seen by 15 million Japanese.[110]

The USIA also targeted the intellectual class in the academy and helped to infiltrate and influence many faculty members and researchers in prominent Japanese universities. As Japanese scholars and professors were underpaid and many had to engage in extracurricular activities to supplement their income, the agency entered into many collaborative financial and intellectual arrangements with them. For example, when left-leaning scholars began to dominate Kyoto University in the early 1950s, then-president Shunjiro Hattori met with USIA staff in 1952 to express his concerns regarding the radical scholarly infiltration.[111] The USIA then sent five young anti-leftist faculty members to the U.S. to learn about industrial development and strengthen their anti-Marxist and pro-American ideologies, so that they could become pioneers of conservative scholarly movements in the university.[112] They eventually became deans of faculty and used their influence to denounce Marxist and pro-communist faculty members within the university. Other anti-left professors, who were identified and contacted by the USIA, were also invited to give lectures at the United Nations Association of Japan (UNAJ), which was supplied with research materials and funds by the USIA.[113]

Meanwhile, the Japanese government passed the law to fund a nuclear research program on March 4, 1954, and passed the Atomic Energy Basic Law (AEBL) on December 16, 1955, which established the Japan Atomic Energy Commission (JAEC).[114] The AEBL codified Japan's intention to utilize atomic energy for the "improvement of the welfare of human society and of the national living standard," thereby paving the legal foundations for the establishment of the JAEC.[115] Shoriki, still the CIA's key asset who became the most avid media cheerleader of nuclear energy, then became the first director of the Science and Technology Agency (STA) and the first chairman of the JAEC. In 1957, he joined the first Kishi Cabinet as chairman of the National Public Safety Commission, while still maintaining his JAEC and STA portfolios.

Soon after, Japan purchased its first commercial reactor from the General Electric Co. (GEC) in Great Britain and built its first nuclear plant at Tokaimura, Ibaragi Prefecture. Japan then quickly switched to U.S.-designed light-water reactors. By mid-1957, the Japanese government entered into a contract to purchase twenty additional reactors from the U.S.[116] After observing the successful transformation of Japanese attitudes and government compliance with the U.S. led-campaigns, the U.S.-led nuclear industries sought to advance its commercial interests through an active campaign of public education at the Tokai Japan Power Demonstration Reactor (JPDR) in Tokaimura.[117]

CIA OPERATIVE DANIEL STANLEY WATSON AND CIA INFORMANT HIDETOSHI SHIBATA

One particular relationship must be reviewed and scrutinized in order to examine the extent of American infiltration into the inner circle of Japanese power brokers.

CIA informant Hidetoshi Shibata indicated in his memo that the Japanese in general had been chronically suffering from so-called an "atomic allergy" or "allergic reactions" to nuclear power. He contended that such a deep-rooted fear and anger spread like a huge mushroom cloud of atomic bombs exploded in Hiroshima and Nagasaki.[118]

He cultivated an extensive network with Prime Minister Shigeru Yoshida, many of his high-ranking officials, and key U.S. officers in American intelligence agencies both in Japan and the U.S. During the period of the Allied forces military occupation of Japan, Shibata also developed and maintained communication links with high-ranking officers of the GHQ.[119]

Shibata frequently met Daniel Stanley Watson, a CIA operative, at the sushi restaurant called Minamoto in Tokyo.[120] Watson traveled to Japan in June 1953 and met Shibata through a British reporter who resided in Tokyo working for the UK's *Sunday Times*, with the objective to get close to Shoriki, who then was the successful owner and president of the *Yomiuri Shimbun*.[121] Access to the media mogul was particularly important for Watson, who understood that "the Japanese read at least three different newspapers daily, then began to construct their opinions and perspectives. *Yomiuri Shimbun* had a largest army of obedient workers underneath the control and commanding structure led by Shoriki. In order to carry out my mission, the success depended upon the extent of communication and influence that I could exert on Shoriki-san."[122]

Watson's objective was to meet a goal specified in the National Security Council (NSC) Progress Report, claiming that "the violence of Japanese reactions to any matter related to nuclear weapons is an element in all of our relations with Japan and raises particular problems in connection with any further U.S. tests in the Pacific as well as in relation to U.S. actions in the development of peaceful use of nuclear energy. . . . We must review and institute a new psychological strategy program for Japan."[123]

The Soviets also saw this as an opportunity to influence Japan's internal politics by utilizing anti-U.S. sentiments in the Japanese government and the general citizenry. Richard Finn, Director of the Office of Japanese Affairs in the State Department, warned about the danger of left-leaning politicians and activists taking over the Japanese polity altogether, stating that Prime Minister Shigeru Yoshida and his party would no longer withstand the anti-U.S. sentiments and had already lost its control in Japan, while the Left gained more power in galvanizing the public under the rhetoric of anti-U.S. influence.[124]

The meetings between Watson and Shibata became more frequent, especially after the Bikini nuclear blast and its subsequent public revelation of Japanese fishermen's exposure to radiation.[125] Watson did not disclose his background to Shibata, though his name appeared many times in Shibata's personal memos. It was only known to Shibata that Watson was involved in the psychological strategies of the U.S. intelligence agencies.[126]

A recent declassified CIA document revealed that Watson also kept detailed notes on Shibata and reported them to his superiors, indicating that "Shibata recognized [him] as a secret intelligence officer, but [did] not know that [he was] a CIA operative."[127] His name, "Stanley Daniel Watson," also appeared in 1996 in another CIA declassified document, *JFK Assassination Records Collection: Lee Harvey Oswald, the CIA, and Mexico City*, revealing that in 1963, Watson was also working as the assistant director of the CIA branch office in Mexico City, monitoring the activities of Lee Harvey Oswald in Mexico.[128]

In the mid-1980s, at Mexico's prominent resort city Puerto Vallarta, Daniel Stanley Watson finally agreed to be interviewed by a reporter from the Nihon Hoso Kyokai (Japan Broadcasting Corporation or NHK). Watson disclosed that he worked in many places in the world, including Tokyo, Hong Kong, and Pakistan. During the interview, he maintained that he never disclosed his true background or connection to American intelligence agencies, or even legitimate reasons for his stay in Japan. He merely stated that he did not tell anyone, even Shibata, about his institutional affiliation or the real purpose of his visit to Japan. "That's all I can say," Watson responded to the reporter.[129] Nonetheless, Watson acknowledged that he approached Shibata because Shibata specifically served as a personal secretary to Shoriki and also had already established a close connection to the prime minister's office and had access to internal information from his office. Watson stated, "I was surprised that information [given by Shibata] was not the kind of information shared among people like TV stations' CEOs, a very quite different level of information."[130]

CIA COLLABORATORS—SHORIKI, SHIBATA, & WATSON

With the support of U.S. intelligence agencies, Shoriki launched a venture involving TV networks (Nihon Terebi Hosomo Kabushiki Gaisha, Nippon Television Network Corporation, or NTV) and began broadcasting in August, 1953, with the intent to create the collaborative network with newspapers and TV stations in Tokyo and, later, other media outlets in the rest of Japan.[131] The NTV became the first commercial TV network in Japan, with the exception of the NHK, which was originally founded in 1925 by the Japanese government as Japan's first national public broadcasting organization.[132] It began its first general TV broadcast in 1953.[133]

Shoriki had long recognized the importance of the mass media and its power in disseminating political messages and influencing public opinion, while making tremendous profits through the advertisement and sponsorship of politically produced shows and programs. His interest in extending his power over mass media outlets was evidenced in his attempt to begin radio broadcasts before the end of World War II. Shoriki first increased the circulation of his paper by making his *Yomiuri Shimbun* the first newspaper to print the scheduled radio program in the paper. Later he attempted to open up the radio station, but his effort failed due to political opposition from the Japanese governmental bureaucracy.[134]

Watson indicated that, with Shibata's assistance, he was able to communicate with Shoriki, even before the 1954 Fukuryu-maru incident, noting that "Shoriki was sharp, asking all the right questions, and [he] just went right to the point on the peaceful utilization of Atomic energy. Japan is the ideal country in the world to make an appeal because it is the only industrial country without much domestic, natural resources. Then I saw Shoriki's eye flashed."[135]

Shibata provided the plan to Watson to invite to Japan a select group of prominent American scientists and business leaders as the ambassador and representative of the Atoms for Peace program from the U.S. Watson also indicated to Shibata that, with sufficient funding, Watson would help organize the tour with the assistance from the U.S. government and private corporations, including General Dynamics.[136]

On January 21, 1954, nuclear submarine USS *Nautilus* (SSN-571), produced by General Dynamics was successfully launched in the U.S. for the first time. GD president John Hopkins was contacted by Shibata, who then asked him to come to Japan as a member of the Atoms for Peace ambassador group and promote the peaceful utilization of atomic energy.[137] On January 1, 1955, *Yomiuri Shimbun* ran the top news on the U.S. ambassador of the Atoms for Peace program, printing news articles and organizing events and special programs to promote the peaceful utilization of atomic energy for the next six months.[138]

PROPAGANDA CAMPAIGNS AND THE ALTERATION OF PUBLIC OPINION

On January 4, 1955, the political matters involving the Fukuryu-maru incident were finally closed after the U.S. decided to pay two million dollars as financial restitution to the Japanese government. The commissioner of the U.S. Atomic Energy Commission, John A. Hall, who also advised President Eisenhower in making his Atoms for Peace speech to the UN in 1953, searched for the proper tone in negotiation with Japanese government on the Fukuryu-maru incident.[139] He suggested that the agreement should have a specific clause, specifying that Japan would not seek the U.S. government's liability on the matter. He also suggested that any information on the bilateral nuclear agreement for the transshipment of uranium from the U.S. to Japan should not be disclosed while the Fukuryu-maru investigation was still underway. Thus, the signing of the official nuclear agreement was delayed after the Fukuryu-maru investigation was completed. On January 11, exactly one week after the Fukushu-maru case was officially closed, the U.S. government sent the letter to the Japanese government that it had accepted the Japanese request to transship the uranium to Japan, but the governmental communication was kept secret from the public.[140]

Despite the propaganda campaigns with the extensive use of mass media, many people still remained highly skeptical of nuclear technology in general. While *Yomiuri Shimbun* had the largest circulation among Japanese newspapers, other news sources continued to disseminate the skepticism and concerns about the use of atomic power. For example, on April 14, 1954, soon after the Fukuryu-maru incident, *Asahi* newspaper ran a big story on the U.S. government's proposal to transmit part of the proprietary knowledge of nuclear technology to the Japanese government.[141] Many people expressed deep concern and some were even outraged, insisting that Japan should not become subservient to U.S. policies in nuclear technology, while others suggested that Japan had to develop its own technology without relying on the U.S. government and its allied corporations.[142]

Prominent Japanese physicist Mitsuo Taketani also opposed Japan's import of the U.S. enriched uranium and nuclear technology, insisting that there was no clear separation of military uses of nuclear power from the commercial application of atomic energy. He was also a prominent Marxist, having published the *Doctrine of the Three Stages of Scientific Development*, in which he elaborated the philosophy of science with Hegel's theory of dialectics.[143] Nonetheless, he gradually began to show less resistance to the propaganda that atomic energy was a "wonderful thing," and believed that the "leftover" from the military research activities could provide the important foundation for the commercial application of atomic energy in Japan.[144]

Still many Marxist scholars and left-leaning scientists were not convinced and remained critical of the potential application of atomic energy for military purposes. In close collaboration with the CIA in Japan, Shibata then produced a detailed report in 1955 examining the political attitudes of prominent scientists, who worked either directly or indirectly on the development of nuclear energy programs in Japan. These investigations were carried out with the help of Koanchosachou (Public Security Investigation Agency) and Keisatusho (National Police Agency), which was given the CIA cryptonym of PODAUB.[145] University professors, researchers, scientists, and scholars who were identified to be anti-government, left-leaning, or antinuclear were marked with a red circle in Shibata's notes.[146]

SHORIKI AND THE INTERNAL POLITICS WITHIN THE JAPANESE GOVERNMENT

In February 1955, Shoriki decided to run for public office in Toyama's second district in order to achieve the following two objectives: (1) to create and stabilize the political groups supportive of the nuclear program, and (2) then to successfully propagate the peaceful utilization of atomic energy through his government office and political network.. He easily won the election and, on April 28, he created his own political vehicle in support of the nuclear program, under the title of the Peaceful Use of the Atomic Energy Committee (Gensiryoku Heiwariyou Kondankai or PUAEC) and served as its chairman and organizer. On the committee were powerful economic political elites, including Ichiro Ishikawa, the president of Keidanren (the Japan Federation of Economic Organizations), along with many prominent scholars and scientists specializing in nuclear research.[147] CIA operative Daniel Watson said that without Shoriki's support, there was no way to assemble such a powerful group of elites from both political and academic spheres, adding that many reluctant scientists were also forced to become its members because of enormous political pressures placed on them.[148]

Only ten days after the establishment of the PUAEC, the Atoms for Peace ambassador arrived in Tokyo on May 9, and his *Yomiuri Shimbun* began to run major stories on the events and activities of the U.S. ambassadors in Japan. *Yomiuri Shimbun* was the major sponsor of the delegate from the U.S., which included Genegal Dynamics president John Hopkins as an Atoms for Peace ambassador, along with Nobel laureate Ernest O. Lawrence and physicist Lawrence R. Hafstad, who worked in the development of the nuclear-powered submarine *Nautilus*.[149] They met Prime Minister Hatoyama and other key political members to urge the signing the U.S.-Japan Atomic Energy Agreement (Nichibei Genshiryoku Kyotei), the precondition for the shipment of enriched uranium from the U.S. to Japan.[150]

On May 13, at Hibiya Kokaido Public Hall for a lecture on the peaceful utilization of atomic energy (Genshiryoku Heiwariyo Dai-koenkai), the delegate made a strong appeal to the Japanese citizenry.[151] Many who failed to enter the auditorium watched and listened to the television screen outside the auditorium. During parts of lecture, prerecorded U.S. commercials were shown to teach the Japanese audience about the safety and application of atomic energy in medical and scientific areas, while purposely excluding the discussion of its potential application in military research and technologies.[152]

The successful public lecture was evidenced in Shibata's personal memo, citing that it was such a great success that the *Yomiuri Shimbun* spent two pages reporting on the entirety of its contents of the presentation.[153] The NTV station cancelled scheduled programs and ran the lecture and presentation live throughout its affiliated networks in Japan.[154] The delegates also traveled to and conducted presentations in Kyoto, Hiroshima, and four other cities.

One principal message hidden behind the delegation's lecture tour, nuclear exhibits, and its massive propaganda campaign through the media was a popular Japanese adage of "Dokuniwa Doku o Motte Seisuru" (control poison with poison, or poison drives out poison). Despite the strong animosity against atomic power shared among the Japanese, this allergic reaction to atomic energy was instead used to drive out or even eliminate its negative effect by embracing the positive use of nuclear energy for peaceful purposes, especially in a country where it has been repeatedly emphasized by the government that it has little or no natural resources.

Shibata's memo substantiated that, while witnessing the these transformative events backstage, he "rejoiced at the successful impact of powerful messages in altering the perception of many skeptical Japanese people."[155] The CIA report dated May 20 which was later wired to the CIA headquarters in the U.S. indicates, "The trip and Hopkins' official statement had received quite positive reactions. For the first time, we were able to steer public attention toward our programs that focused on the peaceful utilization of nuclear energy. No [Japanese] newspaper [for example] covered our first-ever nuclear experiment in a seabed off the coast of California."[156]

On May 19, soon after the live lectures were broadcast by TV stations throughout Japan, the Atomic Power Utilization Preliminary Investigative Committee (Genshiryokuriyo Junbi Chosakai, or APUPIC), which was established within the Economic Planning Agency (Keizai Kikakucho) of the Japanese government, finally announced that it accepted the U.S. offer of enriched uranium to be used for Japan's nuclear program.[157] This was the moment when both the political objectives of Shoriki and the private interests of assembled members of commercial sectors assembled by Shoriki finally joined.

As the first chairman of the APUPIC, Takehisa Shimamura said that the nuclear agreement also had an alternative objective of putting Japan under the mantle of U.S. protection, ensuring that future Japanese action should not deviate from what was expected from the U.S. government's projection of Japan's political role and place in Cold War politics.[158] On June 21, both governments officially signed the U.S.-Japan Atomic Energy Agreement in Washington, D.C., only one year and three months after the Fukuryu-maru incident, finalizing the governmental agreement on the transshipment of enriched uranium from the U.S. to Japan.[159]

Half a year later, Shoriki was appointed as the first Nuclear Energy Minister in the Ichiro Hatoyama cabinet. On November 17, Shoriki sent a personal letter to President Eisenhower, thanking him for the successful tour of the delegation in Japan. The letter specifically indicated that, "[the] Delegation headed by Mr. Hopkins and Drs. Lawrence and Hafstad, marked a turning point in the controversial issue of atomic energy. This movement was the direct stimulus for the conclusion by our government[s] of an atomic energy agreement." [Signed as the Owner, *Yomiuri Newspaper*].[160]

The success of the propaganda campaigns was also recognized by the U.S. government in its report entitled "the National Security Council's Joint Progress Report, State Department and AEC, on implementation of NSC 5507/2—'Peaceful Uses of Atomic Energy.'"[161] It was the summary report on the impact of the Atoms for Peace program, indicating that "during the last three years of intense propaganda against nuclear weapons, and especially against testing, the U.S. position has been more acceptable throughout the Free World because of the parallel efforts to spread the benefits of atomic energy. Although it cannot be measured, this accomplishment of the Atoms-for-Peace programs is substantial."[162]

Meanwhile, Shoriki's JAEC officially approved the APUPIC's plan to purchase a research type 50kW Boiling Water Reactor (BWR, or JRR-1) fueled by enriched uranium solution from the U.S. in 1956, and a CP-5 Heavy Water Reactor (HWR, or JRR-2) fueled by enriched uranium from the U.S. in 1956. The plan also stated that another 10MW HWR (JRR-3) using natural uranium was to be installed in 1958. All three nuclear research reactors were built at the Japan Atomic Energy Research Institute (JAERI) in the remote village of Tokai, Ibaragi Prefecture.[163]

After the enriched uranium finally arrived in Japan, the JRR-1 nuclear power plant first succeeded in reaching nuclear fusion on August 27, 1957, followed by JRR-2 in October 1960, and finally JRR-3 in September 1962.[164] The JRR-3 also became the first domestically built nuclear reactor. But it took another ten years from the JRR-1's first successful fusion to finalize Japan's actual commercialization of nuclear electricity in 1966.[165]

Meanwhile, the U.S. government proposed to the UN General Assembly the creation of an international organization with the primary function of evaluating and seeking the peaceful use of nuclear energy. The U.S. then helped to draft the International Atomic Energy Agency (IAEA) statue, and eighty-one member states of the UN unanimously approved it in October 1956.[166] In July 1957, the IAEA was officially established with its headquarters in Vienna, Austria.[167]

By 1958, the U.S. government had signed the atomic energy agreement with thirty-nine countries, preventing other nations from using nuclear energy for military purposes.[168] The establishment of the IAEA and bilateral governmental agreements with the U.S. had placed other countries under the grand design of Cold War power politics and legally allowed the U.S. and other powerful states to own nuclear weapons, while legitimizing official reviews and inspections of nuclear programs in other countries.

THE CIA AND THE JAPANESE CABINET

Besides Hidetoshi Shibata, who served as Shoriki's personal secretary, other CIA operatives and infiltrators were recruited to work deep inside the Japanese polity.

For example, a senior military adviser to Prime Minister Shigeru Yoshida in the 1950s was an informant for the CIA.[169] Lt. General Eiichi Tatsumi, who was given the CIA code name of POLESTAR-5, advised the prime minister on defense-related matters.[170]

Fresh from the memory of nightmarish devastation at home following World War II, Prime Minister Yoshida refused to consider the possible remilitarization of the Japanese armed forces, while the U.S. government that had just concluded the military conflict in the Korean Peninsula in the early fifties wanted Japan to remilitarize against the potential communist threat in the region. Tatsumi closely and actively communicated with the CIA in providing crucial information on internal governmental positions and debates about possible remilitarization.[171]

Lt. Gen. Tatsumi was also a key member of an anti-communist Japanese spy organization called Kawabe, which was established at the request of the GHQ (General Headquarters) during the U.S.-led occupation of Japan and later closely collaborated with the CIA. Kawabe was active during the execution of its clandestine operations and activities, including the formulation of a plot to assassinate Prime Minister Yoshida, though the plot was not carried out.[172] The CIA file on Tatsumi from 1952 to 1957 substantiated that he was actively involved in the establishment of the new military and intelligence organizations in Japan.[173] In a CIA memorandum dated November 26, 1956, he was portrayed as "probably one of the best, safest, most qualified persons in Japan today for CIA use."[174]

Prominent Japanese politician Taketora Ogata also played the role of another key informant for the CIA, which gave him the cryptonym of POCAPON.[175] A recently declassified five-volume-, 1,000-page document collection called "the Ogata File" revealed the content of the detailed CIA efforts to support election campaigns of anti-communist conservative politician Taketora Ogata, who then led the Japan Liberal Party in the early 1950s and played a pivotal role in merging his party with other conservative political factions to form the Liberal Democratic Party (LDP).[176] Shoriki's *Yomiuri Shimbun* then became the most influential media supporter of the Liberal Democratic Party.[177] The CIA identified Ogata as the most useful "asset" among the leaders of Japanese conservative politicians. Although Ogata passed away in 1956, his LDP party, with active CIA assistance, went on to rule the Japanese government almost without interruption between 1955 and 2009, effectively transforming the country into a one-party state.[178]

SHORIKI AND THE END OF HIS COLLABORATION WITH THE CIA

It is clear that Shoriki viewed the CIA and its clandestine operation as a useful channel for obtaining U.S. support to pursue and achieve his ambitious economic and political objectives. As a former prewar military bureaucrat, a post-war prominent business leader, and a powerful political power broker, the CIA's approach to, and collaboration with, Shoriki allowed him to cultivate a crucial connection and access to powerful American policy makers in order to seek Washington's support for a grand scheme of developing a nationwide microwave communication network for his private media enterprise.[179] Shoriki ultimately sought to become prime minister and viewed the development of Japan's successful nuclear program as a political springboard to attain his ultimate political goal.[180]

As his relationship with CIA deepened, Shoriki began to use the CIA as a back door to influence Washington on a number of occasions. For example, he demanded that the U.S. government change its planned site for the establishment of the Asian Nuclear Center from the Philippines to Japan.[181] He also demanded, without delay, the export of an American nuclear reactor to Japan.[182] The CIA document dated December 28, 1955, explained that a memo provided by CIA informant Shibata stated that Shoriki was irritated by the U.S. insistence on building the Asian Nuclear Center in Philippines and felt the necessity to persuade the U.S. to relocate it in Japan in order for Japan to become a leading anti-communist country in Asia.[183] The CIA document of the following day also stated that "as our recent reports indicate, the PODAM [Shoriki] is becoming extremely ambitious. As his political role in the government has enlarged enormously, his capacity to carry out our mission has also been enlarged. . . . His media empire is the most centralized in Japan . . . and perhaps the most efficient in the propaganda campaign to capture the hearts and minds of the Japanese public, which is something any politicians are eager to capitalize."[184]

In order to obtain the proprietary nuclear technology and bring the American nuclear reactor to Japan, Shoriki even went beyond the CIA and directly contacted General Dynamics president Hopkins, asking him to intervene in the governmental decision-making process to streamline the transportation of the nuclear power plant to Japan.[185] The CIA report cited that "the PODAM even concocted the private scheme to have Japanese Ambassador [Sadao] Imai visit the U.S. government next week to ask for a decision favorable to Japan."[186]

The CIA document dated December 2 also indicated that they were forced to terminate another CIA operation called the DISEM because of the insider's request to veer away the PODAM's further demands and prevent the realization of his ambitious plan.[187] The DISEM refers to the CIA operative's project to establish and construct the strategic microwave telecommunication network in Japan and later globally. A week later, on December 9, the CIA report indicated that "our collaborative project resumes to provide an indirect hand to the restoration of Japan's influence. Ultimate goals of this man's [PODAM's] ambitions cannot be overlooked. One example will be the construction of the microwave telecommunication network. If this would be completed, Japan could have an ultimate propaganda machinery to influence all other free Asian countries. If we accept his demand on the nuclear energy project in Japan, it, too, can assist Japan in developing its own nuclear weapon. These could maximize the potential [of Japan] to become a serious trouble-maker and provide a crucial tool [for Japan] to attain the first class nation status in the world."[188]

The CIA was fully aware of Shoriki's ambitious goals and ultimate motives. Previously in August, the CIA warned in its report that Shoriki was a strategic and shrewd operator.[189] It was well aware of the fact that no conscionable or well-intended Japanese person would fully cooperate and work with the U.S. intelligence agencies. Besides Shoriki, other Japanese who approached these agencies had their own motives as well. Though U.S. intelligence agencies did little to deter them, they wondered if they could control Shoriki by responding to his ever-increasing demands.[190]

As the CIA had anticipated, Shoriki's ambition and demands began to accelerate and his objectives grew even more ambitious. Four days after Shoriki became the first chairman of the JAEC on January 1, 1956, he held the first committee meeting and announced the two top agendas to be accomplished by the Japanese commission: (1) to establish self-sufficiency in the domestic production of nuclear-powered electricity within five years, instead of the previously projected ten, and (2) to sign a bilateral agreement with the U.S. government to build a nuclear power plant and import any other proprietary technology and nuclear know-how from the U.S.[191] The transfer of such proprietary technical knowledge, as Shoriki assumed, included the enrichment of uranium, which the U.S. government was reluctant to provide because it could enable Japan to build its own nuclear weapon.[192] Nonetheless, for Shoriki, the construction of a nuclear power plant in Japan and the commercial production of electric power plants within five years became the foundation of his political slogan, upon which his ultimate ambition to become the prime minister was based.[193] Many prominent Japanese scientists, including Nobel laureate physicist Hideki Yukawa, began to criticize and became skeptical of Shoriki's ambitious political objectives that placed a greater emphasis on the commercial application of nuclear technology and atomic energy, rather than the scientific research and development of the peaceful use of atomic energy in Japan.[194]

In March 1956, Shoriki decided to invite Geoffrey W. Lloyd, British Minister of Brandstoffen Energy to Japan. He successfully helped build the first commercial nuclear power station in Calder Hall, Britain, and delivered electricity in commercial quantities for the first time in the West. Shoriki's stated intention to purchase a nuclear power plant from Great Britain had forced the CIA to reconsider their strategy in dealing with Shoriki and his ever increasing demands.[195]

The CIA report dated on May 5 indicated that "following the decisions made at the April 26 meeting, both the AEC [Atomic Energy Commission] and the Department of State reached the consensus to provide necessary support to Shoriki [and his demands]," including the Department of State's decision to provide a U.S. loan to Japan, as requested by Shoriki and his Japanese Atomic Energy Commission (JAEC), as well as the first export of a nuclear power plant to Japan.[196] The same CIA document also warned that the U.S. decisions and CIA operations should look as if everything was being initiated by the Japanese side and that the U.S. was merely responding to the Japanese demands.[197]

While the U.S. considered the Japanese government as a key ally in building a strong anti-communist network in Asia during the early phase of the Cold War, the CIA operations made sure that the Japanese government's decision to put themselves under the influence of U.S. military protection was their own, not something purposely initiated or unilaterally demanded by the U.S. side. At the same time, the careful reading of CIA documents also revealed that the U.S. was willing to comply with Shoriki's demands in order to maximize the successful execution of their psychological warfare operations in Japan. Nevertheless, it is important to note that the CIA documents do not indicate American's willingness to transfer proprietary nuclear technology to Japan, including the enrichment of uranium and/or the extraction of plutonium.[198]

Recognizing the nuanced reluctance of the CIA to meet his demands or political objectives, Shoriki began to turn away from the CIA and U.S. influences. Instead, he then tried to cultivate and build a new British alliance and collaboration, announcing that the Japanese government would import the nuclear power plant from Great Britain, not the U.S.[199] Given the U.S. reluctance, the first reactor that Shoriki was responsible for constructing at Tokai became not the U.S. nuclear reactor, but the Calder Hall type imported from Great Britain, which was originally developed for the production of plutonium to build nuclear weapons.[200] The utilization of heated steams to spin turbines to generate electricity was only secondary in Shoriki's mind, as he was primarily interested in the acquisition of plutonium, which is the byproduct of the burning uranium.

Shoriki began to be critical of the CIA's activities in Japan and began to use his media empire to criticize and attack U.S. foreign policies and operations. For example, on June 20, 1956, *Yomiuri Shimbun* ran a critical article on its front page that two American soldiers accidentally lost their eyesight because of American nuclear scientists' utter failures to detonate the hydrogen bomb in its designated destination by four miles in the Pacific.[201] *Yomiuri Shimbun* also sarcastically framed the imperialist projection of the U.S. policies in Okinawa, suggesting "while Japan may need a much imposed American pretence of the kindness for her protection, America should not supply Japan's Self-Defense Forces a kind of airplanes that could easily catch fire and crash to the ground. The sort of 'kindness' that goes a way off the mark. Such a 'kindness' is the reason that could lose the American popularity in the world today."[202]

Shoriki's continuous demands led to the CIA's decision to critically assess future collaboration with Shoriki. When the U.S. government finally decided not to agree to his demands, Shoriki became critical of the CIA and began to use his media empire to attack U.S. foreign policies in Japan and Asia. The CIA began to be weary of his critical articles and editorials in *Yomiuri Shimbun*, questioning the legitimacy of U.S. foreign policies in Japan and the rest of Asia.[203]

In one instance, after the CIA realized that Shoriki was reluctant to influence his reporters in *Yomiuri Shimbun* and refused to change the contents of anti-U.S. articles in its editorials, it became irritated and its operatives tried to find out the names of reporters and identify their residences for future intimidation and retaliation.[204] The CIA even considered that they should turn Shibata, Shoriki's private secretary and CIA informant, into an independent "asset" without Shoriki's knowledge.[205]

CONCLUSIONS

The CIA and U.S. intelligence agencies worked closely with Shoriki to engage in psychological warfare and advance the campaign to promote the peaceful utilization of nuclear energy in Japan. The close collaboration, though at times contentious, was well-founded because Shoriki provided a valuable resource on numerous fronts, both through his connections to internal politics and economics, as well as his authority to mobilize Japanese mass media. He also proved a natural "asset" for the CIA's psychological warfare strategies, as his deep insights into the importance of mass media and the effectiveness of propaganda campaigns allowed him to effectively utilize them in the pre-war era for advancing Japanese imperialist projects until 1945.

The CIA's own propaganda campaign proved so effective that even today, some of the atomic bomb victims in Hiroshima failed to see through its ideological veil. Eighty-three-year-old former Hiroshima City Major Takashi Hiraoka recently divulged in a radio interview that, until the Fukushima nuclear disaster happened, he remained a strong advocate of the use of atomic energy and was incapable of identifying commonalities between peaceful use and military applications of atomic power.[206] He indicated that his distorted belief was cultivated when he visited the Atoms for Peace exhibition in Hiroshima as an impressionable young newspaper reporter working for the local *Chugoku Shimbun*.[207]

Despite the successful campaign and the long collaboration between the American intelligence agencies and Shoriki, it became apparent that the CIA and its affiliates in the State Department found it increasingly difficult to control Shoriki, whose ultimate aim was fixated with the ambition to import nuclear technology necessary to engineer Japan's rearmament process, while personally pursuing his goal to become Japan's prime minister on the foundation of building Japan's successful nuclear program.

The CIA documents clearly showed that the U.S. government had no intention of equipping its former wartime enemy with proprietary technology and making Japan capable of building its own nuclear weapon. Eisenhower's Arms for Peace policy was rather to make the Japanese public and their government compliant with American nuclear foreign policies, while seeking to expand the future market for nuclear technology developed by U.S. corporations.

As in all psychological propaganda campaigns, it is impossible to measure the extent of genuine impacts and effects precisely. Nonetheless, banking on the nuclear safety myth, the Japanese government has promoted an aggressive policy of locating and building nuclear power plants in depopulated areas suffering from a significant decline of local industries and economies. The Fukushima nuclear disaster thus substantiated that U.S. propaganda programs left a long-lasting legacy in Japan. The U.S. manufacture of consent on the safety of nuclear energy, supported by Japan's news media, has created fertile ground for a future nuclear disaster. Even after Fukushima reactors exploded and massive radiation fallout contaminated many regions, the Japanese government and corporate media began to engage in another propaganda campaign to create a new kind of the myth—"low-level radiation is safe," especially in Fukushima where more than millions of people were still trapped in radiation-contaminated areas. Whether or not the Japanese people will liberate themselves from the past legacy of the U.S.-imposed propaganda and nuclear programs and discontinue relying on atomic energy to find their own democratic path toward equality, equity, and safety is something that the Japanese people alone must decide in coming years.

The remaining chapters of this book offer a critical analysis of the impact of Japan's nuclear policies and disaster in the past, present, and future.

NOTES

This chapter was presented at the Law & Society Association Conference in Boston, May 2013. We thank William Qiu at Stanford University for his critical comments and suggestions. Finally, translations are our own unless otherwise provided.

1. Postwar CIA psychological warfare strategies and propaganda activities in Japan were critically analyzed by Waseda University Sociology Professor Tetsuo Arima, using recently declassified CIA documents. His critical analysis of the CIA and U.S. intelligence agencies and their operations in Japan was published in the following three books: (1) *Genpatsu, Shoriki, CIA: Kimitsu Bunsho de Yomu Showashi [Nuclear Power, Shoriki, & CIA: History of Showa from Secret Documents]* (hereinafter Arima-2008), (2008); (2) *Showashi o Ugokashita Amerika Johokikan [American Intelligence Agencies that Influence the History of Showa]* (hereinafter Arima-2009) (2009); and (3) *CIA to Sengo Nihon* [CIA and Postwar Japan] (hereinafter Arima-2010)(2010).

2. David McNeill, Japan Must Develop Nuclear Weapons, Warns Tokyo Governor, the *Independent*, March 8, 2011, available at http://www.independent.co.uk/news/world/asia/japan-must-develop-nuclear-weapons-warns-tokyo-governor-2235186.html.

3. The Soviet Union, not Russia, first engaged in nuclear testing in 1949, followed by the United Kingdom in 1952. France first engaged in nuclear testing in 1960.

4. The Japanese government finally passed the Atomic Energy Basic Act (AEBA) in December 1955, establishing the Japanese Atomic Energy Commission (JAEC). The Chapter 1 "General Provision" of the AEBA indicates "to secure energy resources in the future and achieve the progress of science and technology and the promotion of industries by encouraging the research, development, and utilization of nuclear energy, and thereby contributing to the improvement of the welfare of human society and of the national living standard." See www.nsc.go.jp/NSCenglish/documents/ laws /1.pdf.

5. Walter E. Grunden, Mark Walker, and Masakatsu Yamazaki, Wartime Nuclear Weapons Research in Germany and Japan, *OSIRIS*, 20:107-130, 2005.

6. See Peter Kuznick, Japan's Nuclear History in Perspective: Eisenhower and Atoms for War and Peace, *Bulletin of the Atomic Scientists*, April 13, 2011, available at http://www.thebulletin.org/web-edition/features/japans-nuclear-history-perspective-eisenhower-and-atoms-war-and-peace; From Hiroshima to Fukushima: The Political Background to the Nuclear Disaster in Japan, *World Socialist Web Site*, June 23, 2011, available at http://wsws.org/articles/2011/jun2011/fuku-j23.shtml.

7. Memorandum from the Director of the Foreign Operations Administration (Stassen) to the Duty Assistant to the President (Persons), June 27, 1955, FRUS, 1955–1957, Vol. X: p. 11.

8. Japan's Nuclear Power Policies, JapanEco.Net, May 26, 2011, available at http://japanecho.net/311-data/1019/.

9. Ex-Bureaucrat Details Secret U.S. Nuke Pact, *Japan Times*, June 30, 2009, available at http://search.japantimes.co.jp/cgi-bin/nn20090630a2.html.

10. Kakuo Motometa Nihon [Japan that Sought Nuclear Weapon], NHK Scoop-Document, October 3, 2010.

11. Id.

12. Philip Brasor, Nuclear Policy was Once Sold by Japan's Media, *Japan Times*, May 22, 2011, available at http://search.japantimes.co.jp/cgi-bin/fd20110522pb.html; see also Arima-2008, supra note 2.

13. National Archives, Research Aid: Cryptonyms and Terms in Declassified CIA Files: Nazi War Crimes and Japanese Imperial Government Records Disclosure Acts, June 2007 (hereinafter cryptonym), available at http://www.archives.gov/iwg/declassified-records/rg-263-cia-records/second-release-lexicon.pdf. The key propaganda campaign involved the CIA and Pentagon support for Shoriki to establish Japan's first-ever commercial television network, called the Nippon Television Corporation; See also Arima-2010, supra note 2, at 215.

14. Cryptonym, supra note 14.

15. MIT: Origin of the Center, the MIT Center for International Studies (last visited August 24, 2011), available at http:// web.mit.edu/cis/pdf/Panel_ORIGINS.pdf.

16. Id; Arima-2009, supra note 2, at 193–194.

17. Id.

18. Craig Doyle Nelson, Nuclear Bonds: Atoms for Peace in the Cold War and in the Non-Western World, master's thesis at The Ohio State University (2009), available at http://etd.ohiolink.edu/send-pdf.cgi/Nelson%20Craig%20Doyle.pdf?osu1237397691.

19. Richard Katz, How Japan's Nakasone Should Play His Cards Against Henry Kissinger, 48 *Executive Intelligence Review*, 47, 49 (1982), available at http://www.larouchepub.com/eiw/public/1982/eirv09n48-19821214/index.html.

20. Arima-2008, supra note 2, at 251.

21. See also "Former Prime Minister Nakasone Receives 'Atoms for Peace' Prize," Atoms in Japan: News in Brief, Oct. 29, 2008, available at http://www.jaif.or.jp/english/aij/member/2008/2008-10-29a.pdf . See also, http://www.atoms-for-peace.org/prize/claureat-2008.htm .

22. See generally Yasuhiro Nakasone, *Heaven and Earth Have Feelings: Speaking of Fifty Years of Postwar Politics* (1995).

23. Anthony J. Hall, From Hiroshima to Fukushima, 1945–2011, World Awash in Environmental Armageddon, April 21, 2011, available at http://environmentalarmageddon.wordpress.com/; See Arima-2009, supra note 2, at 200. Shibata later became the CEO of the Nippon Television Network. See Arima-2008, supra note 2, at 58.

24. Hall, supra note 16.

25. See generally Ichizo Kanmatsu, *Nihon Terebi Hosomo Koso o Shoriki Matsutaro [Nippon Television Network Projects and Matsutaro Shoriki]* (2005).

26. Arima-2008, supra note 2, at 24.

27. Arima-2008, supra note 2, at 11.

28. Louise Williams and Roland Rich, *Losing Control: Freedom of the Press in Asia* 97 (2000) ("Shoriki took a leading role in the 'state party' Imperial Rule Assistance Association").

29. Simon Partner, *Assembled in Japan: Electrical Goods and the Making of the Japanese Consumer* (NY: Columbia University Press, 1999), p.74.

30. See generally Mitsutaro Ishii, *Kaiso 88 Nen* [Memoir of 88 Years] (1976).

31. Karel van Wolferen, *The Enigma of Japanese Power* 362 (1990) (Shoriki "exerted himself on behalf of the bureaucrats in the 1936 merger of the news agencies into one government propaganda organ, Domei Tsushinsha."); See also Michio Nagasawa, Senso to Shimbun [War and Newspapers], DIG Newsletter, August 2007, available at http://homepage2.nifty.com/DIG-Japan/senso_to_sinnbunn.htm .

32. Bunzo Hashikawa, Kakushin Kanryo [Reformist Bureaucrats] in Jiro Kamishima (ed.), *Kenryoku no Shiso* (Tokyo, Chikuma Shobo, 1965), p. 256.

33. Joe Moore, *Japanese Workers and the Struggle for Power 1945–1947*, 48–49 (1983).

34. Id., at 49.

35. Stephen C. Mercado, Book Review: A Limited Japanese View of the CIA, *International Journal of Intelligence and Counterintelligence*, 24: 211–216 (2011).

36. Id., at 213.

37. Eric Johnson, Key Players Got Nuclear Ball Rolling, *Japan Times*, July 16, 2011, available at http://search.japantimes.co.jp/cgi-bin/nn20110716f1.html .

38. Arima-2008, supra note 2.

39. Cryptonym, supra note 14.

40. Id.

41. Id.; See also Central Intelligence Agency, Intelligence in New Japan, September 18, 1995, available at https://www.cia.gov/library/center-for-the-study-of-intelligence/kent-csi/vol7no3/html/v07i3a01p_0001.htm .

42. Arima-2008, supra note 2, at 35.

43. Id.

44. For specific individual members of the Japan Lobby, see Arima-2010, supra note 2, at 36–37; See also Hall, supra note 16.

45. Id.; See also Tetsuo Arima, Reisen no Media Nippon Terebi Hosomo: Shoriki, Maikurowe-bumo o Meguru Geikoku Hankyosanshugi Kaiko Johoseisaku [Media in Cold War and

Nippon TV Network: Shoriki & America's Anti-Communism Foreign and Intelligence Policies, available at www.f.waseda.jp/tarima/coldwarNTV.pdf.

46. The Department of Defense Role in Foreign Assistance Background, Major Issues, and Options for Congress, Congressional Research Service, December 9, 2008, available at www.fas.org/sgp/crs/natsec/RL34639.pdf.

47. Chalmers Johnson, *Nemesis: The Last Days of the American Republic* 93 (2006).

48. Arima-2009, supra note 2, at 185–215.

49. Id.

50. See generally Arima-2010, supra note 2, at 124–127 (explaining the CIA's Operation PODALTON to secure $10 million to help establish Japan's first commercial TV network); see Arima-2009, supra note 2, at 216–220.

51. Arima-2008, supra note 2, at 32–35.

52. Jack K. MacFall-Holthusen, TV World wide Network Japan, April 4, 1952, Holthusen Papers, Box 8 in Hubert Hoover Presidential Library, as cited in an article, From Hiroshima to Fukushima: The Political Background to the Nuclear Disaster in Japan, June 23, 2011, World Socialist Web Site, available at http://www.wsws.org/articles/2011/jun2011/fuku-j23.shtml.

53. Id.

54. James Schwoch, Crypto-Convergence, Media, and the Cold War: The Early Globalization of Television Networks in the 1950s, May 2002, available at http://web.mit.edu/cms/Events/mit2/Abstracts/MITSchwochTV.pdf.

55. Id., p. 27.

56. David E. Kaplan, U.S. Propaganda Efforts in Postwar Japan, Japan Policy Research Institute, February, 1997, available at http://www.jpri.org/publications/critiques/critique_IV_1.html

57. Id.

58. Id.

59. Id.

60. Chris Hedges, *Death of the Liberal Class* 69 (2010).

61. Id.

62. Kuznick, supra note 4.

63. Stanley Levey, Nuclear Reactor Urged for Japan T. E. Murray of A.E.C. Tells Steel Union Step is Vital in Atom Race with Russia, *New York Times*, p. 14, September 24, 1954.

64. Schwoch, supra note 55.

65. Id.

66. Tetsuya Arima, Cold War, East Asia, and Television, *Cultural Studies Monthly*, August 25, 2007.

67. Phil Taylor, Glossary of Relevant Terms and Acronyms, The Institute of Communications Studies, University of Leeds, 1987, available at http://ics.leeds.ac.uk/papers/vp01.cfm?outfit=pmt&folder=64&paper=665.

68. Id.

69. Arima-2008, supra note 2, at 50–51.

70. Id., at 8–9.

71. One report indicated that Kuboyama's autopsy confirmed "hepatitis as proximate cause of death but left the role of radiation ambiguous." (see Barton C. Hacker, *Elements of Controversy: The Atomic Energy Commission and Radiation Safety in Nuclear Weapons Testing 1947-1974*, p. 158 (1984).

72. NHK: Gendaishi Sukupu dokyumento: Genpatsu Donyu no Shinario, Reisenka no Tainichhi Genshiryoku Senryaku [NHK Current Scoop Document: Scenario for the Indoctrination of Atomic Energy – Anti-Japanese Nuclear Strategies during the Cold War] (hereinafter NHK-1994), NHK, 1994, available at http://video.google.com/videoplay?docid=-5843883287656171344#.

73. Arima-2008, supra note 2, at 47.

74. Id.

75. See *Astroboy 1: Tape One: Birth of Astroboy; The Monster Machine* (NBC television broadcast, 1963).

76. See Tetsuwan Atomu [Astroboy]: *Tezuka Osamu.net*, available at http://tezukaosamu.net/jp/manga_syllabary_search/291.html.
77. See Takuji Ono, *Egakikaerareta Tetsuwan Atomu [Modified Stories of Astro Boy]* (2008).
78. Arima-2008, supra note 2, at 47.
79. Id.
80. Id., at 54, 249.
81. NHK-1994, supra note 70.
82. Kuznick, supra note 7.
83. Id.
84. Kenneth Osgood, *Total Cold War: Eisenhower's Secret Propaganda Battle at Home and Abroad* (Kansas: University Press of Kansas, 2008), p. 179.
85. Yuki Tanaka and Peter Kuznick, Japan, the Atomic Bomb, and the 'Peaceful Uses of Nuclear Power," the *Asia-Pacific Journal*, vol. 9, issue 18, 1, May 2, 2011, available at http://www.japanfocus.org/-Yuki-TANAKA/3521.
86. Id.
87. Arima-2008, supra note 2, at 250.
88. Allan Lazar, Dan Karlan, and Jeremy Salter, *One Hundred One Most Influential People Who Never Lived: How Characters of Fiction, Myth, Legends, Television, and Movies Have Shaped Our Society, Changed Our Behavior, and Set the Course of History* (2006), p. 267; See also Arima-2008, p.83.
89. Arime-2008, supra note 2, at 86.
90. Id, at 251.
91. Id., at 163.
92. Hiroaki Koide, Genpatsu of Anzen Shinwa o Kangaeru [Critique of the Safety Myth of Nuclear Energy], September 4, 2011, available at http://vimeo.com/28821585.
93. Arima-2008, supra note 2, at 144.
94. IMDb, Walt Disney's Wonderful World of Color: Our Friend the Atom (last visited October 11, 2011), available at http://www.imdb.com/title/tt0151857/plotsummary.
95. Marshall Brain, Blast from the Past: Our Friend the Atom, *Howstuffworks*, April 15, 2010, available at http://blogs.howstuffworks.com/2010/04/15/blast-from-the-past-our-friend-the-atom/.
96. Id.
97. Arima-2008, supra note 2, at 206-207, 218.
98. Id.
99. See generally Jules Verne, *20,000 Leagues Under the Sea* (1995).
100. Brain, supra note 94.
101. Id, at 219.
102. Central Intelligence Agency, A Look Back . . . Robert Carey Broughton: From Walt Disney to War Movies, September 4, 2009, available at https://www.cia.gov/news-information/featured-story-archive/robert-carey-broughton.html.
103. Id.
104. Id.
105. Id.
106. IMDb, "Der Fuehrer's Face" (1942), available at http://www.imdb.com/title/tt0035794/ (last visited October 12, 2011).
107. Craig Doyle Nelson, Nuclear Bonds: Atoms for Peace in the Cold War and in the Non-Western World, Ohio State University, 2009, p.69, available at http://etd.ohiolink.edu/view.cgi/Nelson%20Craig%20Doyle.pdf?osu1237397691.
108. Hiroki Sugita, Postwar Propaganda Campaign: USIS Role Revealed in Japan's Tilt Toward West, Japan Times, November 21, 1997, available at http://search.japantimes.co.jp/cgi-bin/nn20071121f1.html.
109. Id.
110. Id.
111. Id.
112. Id.

113. Id.
114. Arima-2008, supra note 2, at 249–250.
115. Atomic Energy Basic Act (Act No. 186 of 1955), Chapter 1: General Provisions, available at http://www.nsc.go.jp/NSCenglish/documents/laws/1.pdf.
116. Hall, supra note 15.
117. Id.
118. Arima-2008, supra note 2, at 59.
119. NHK-1994, supra note 70.
120. Arima-2008, p. 59.
121. NHK-1994, supra note 70.
122. Id.
123. Id. The name of the report is *National Security Council: Progress Report: United States Objectives and Courses of Action With Respect to Japan*. The photocopy of the report is shown in the video.
124. NHK-1994, supra note 70.
125. Arima-2008, supra note 2, at 58–59.
126. Id.
127. Id., p. 58.
128. Id, p. 58–59. The document information is given in Arima-2008, p. 243. The title of the document is called "JFK Assassination Records Collection. Lee Harvey Oswald, the CIA, and Mexico City. JFK. 104-10067-10191.
129. Id.
130. NHK-1994, supra note 70.
131. Id.
132. See NHK's official website, *NHK Profile*, available at http://www.nhk.or.jp/pr/english/profile/pforile.html (last visited October 12, 2011).
133. Id.
134. Arima-2008, supra note 2, at 12.
135. NHK-1994, supra note 70.
136. Id.
137. Arima-2008, supra note 2, at 72.
138. NHK-1994, supra note 70.
139. Hacker, supra note 70, at 357 (see footnote 194, "J.A. Hall to Nichols, "The Funding Arrangements for Possible Compensation to the Japanese Government on the Fukuryu Maru Case," 22 Apl. 1954; Hall to G.C. Smith, 28 Apr. 1954; State Department TWX 2394 to American Embassy Tokyo, 30 Apr. 1954; AEC 730/1, "Funding Arrangements for Possible Compensation to the Japanese Government on the Fukuryu Maru Case," 4 May 1954); see also Atoms for Peace Awards, 1968, available at www.iaea.org/Publications/Magazines/Bulletin/Bull105/10501401011.pdf.
140. Hacker, supra note 70.
141. NHK-1994, supra note 70 (the actual photo of the newspaper headline is shown in the following web page which transcribed the NHK program, Zamaamiyagare, http://blog.livedoor.jp/amenohimoharenohimo/archives/65733311.html).
142. Id.
143. Mitsuo Taketani, *Taketani Mitsuo Chosaku Shu [Works of Mitsuo Taketani]*, 6 vols, Keiso Shobo, 1968; Mitsuo Taketani, *Taketaniya Mituo Gendaironshu [Collection of Modern Writings by Mitsuo Taketani]*, 7 vols, Keiso Shobo, 1974.
144. Martin Dusinberre, How Japan's Nuclear Industry Got Here, Reuters, March 18, 2011, available at http://blogs.reuters.com/columns/2011/03/18/how-japans-nuclear-industry-got-here/.
145. Cryptonym, supra note 14
146. NHK-1994, supra note 70.
147. Id.
148. Id.
149. Arima-2008, supra note 2, at 82–83.
150. Id.

151. Id.
152. Kuznick, supra note 4; See also NHK-1994, supra note 51.
153. Arima-2008, p.86; See also NHK-1994, supra note 70.
154. NHK-1994, supra note 70.
155. Id.
156. Arima-2008, supra note 2, at 88.
157. Id., at 250.
158. NHK-1994, supra note 70.
159. Arima-2008, supra note 2, at 250.
160. Id., at 130; See NHK-1994, supra note 70 (showing an actual photocopy of Shoriki's letter).
161. NHK-1994, supra note 70.
162. Id (the statement came from the photocopy of the actual content of the NSC report shown in the program).
163. Hitoshi Yoshioka, *A Social History of Science and Technology in Contemporary Japan: Vol. 2: Road to Self-Reliance 1952–1959* (2005), pp. 90-91.
164. Arima-2008, supra note 2, at 225.
165. NHK-1994, supra note 70.
166. International Atomic Energy Agency, A Short History of IAEA, http://www.iaea.org/About/history.html (last visited October 10, 2011).
167. Id.
168. NHK-1994, supra note 70.
169. Documents Show Japan Government Aide was CIA Mole, Intelnews.org, October 5, 2009, available at http://intelligencenews.wordpress.com/2009/10/05/03-86/
170. Cryptonym, supra note 14.
171. Documents Show Japan Government Aide was CIA Mole, Intelnews.org, October 5, 2009, available at http://intelligencenews.wordpress.com/2009/10/05/03-86/ .
172. Postwar Defense Aide was CIA Mole, Japan Times, October 5, 2009, available at http://search.japantimes.co.jp/cgi-bin/nn20091005a2.html .
173. Id.
174. Id.
175. Cryptonym, supra note 14.
176. 20 Seiki Media Kenkyujo Tokubetu Kenkyukai: CIA to Ogata Taketora [20 Century Media Research Institute, Special Study Session: CIA & Taketora Ogata] July 25, 2009, available at members.jcom.home.ne.jp/katote/0907OGATA.pdf .
177. "Toru Shoriki: Japanese Businessman Whose Empire Ranged from Newspapers to Basketballs," *The Times*, August 23, 2011, p. 49.
178. Joseph Fitsanakis, Documents Reveal CIA Meddling in Japanese Elections, Intelnews.org, July 28, 2009, available at http://intelligencenews.wordpress.com/2009/07/28/02-118/#more-2366 .
179. Arima-2008, supra note 2, at.14.
180. Id., at 36–57.
181. Id., at 133.
182. Id., at 133–134.
183. Id., at 132.
184. Id., at 132–133.
185. Id, at 134.
186. Id., at 134–135.
187. Id., at 138.
188. Id., at 137.
189. Id., at 138.
190. Id., at 138–139.
191. Id., at 146–147.
192. Id., at 148–149.
193. Id., at 147.
194. Id., at 148–149.

195. Id., at 159; See also Helge Kragh. *Quantum Generations: A history of Physics in the Twentieth Century* (1999), p.286.
196. Id., at 160.
197. Id., at 160.
198. Id., at 162–163.
199. Id., at 167–168.
200. Peter Hayes, "Should the United States Supply Light Water Reactors to Pyongyang? *Nautilus Pacific Research*, November 16, 1993, available at http://web.archive.org/web/20060307121518/http://www.nautilus.org/archives/pub/ftp/napsnet/papers/hayes1193.txt ("Calder Hall Reactors [were] built to produce plutonium for nuclear weapons.")
201. Arime-2008, supra note 2, at 169.
202. Id., at 169–170 (citing the front page article on June 20th, 1956).
203. Id., at 171.
204. Id., at 171–172.
205. Mercado, supra note 36, at 215 ("The CIA also considered turning Shibata Hidetoshi, his lieutenant and conduit to the local Agency station, into a independent asset").
206. Shusenno Hini Gensiryoku no Heiwariyo o Kangaeru [To Consider "the Atoms for Peace" Program at the Anniversary of the End of WWII], *Tanemaki Journal*, April 15, 2011, available at http://www.youtube.com/watch?v=DCwYZ-xbTjI (Hiraoka later became the President of Chugoku Hoso [Chugoku Broadcast Corp.]).
207. Id.

II

Nuclear Fallout

> You know, plutonium is named after Pluto, the god of Hell. And that's an indication that it's a pretty nasty element. It's in all these reactors. Unit Three was using what's called mixed oxide fuel. So Unit Three had more plutonium than the other units. But all of these reactor fuel pools and the nuclear reactors themselves have plutonium in them.
> When plutonium volatilizes, when it gets hot and turns to a vapor, it can be breathed in. And, of course . . . —it can cause cancer in lungs very, very easily. And the containments, which are designed to contain this plutonium, . . . —have failed, at least in Unit Two. I believe in Unit One and Three, they are leaking, but they probably haven't failed. So, it is likely that volatile plutonium is being released right now. [1]

The last wave in Japan will supercede all others, spelling the doom of the nation's millions and most nonaquatic species. The disaster lies far in time's future; but the beginnings are now seen as Japan suffers a second fall from imperious power and well-being.

Japanese social and production relationships depend on inanimate energy, with nuclear power both the leading force and its fallible weakness.

The fuse igniting nuclear power in turn depends on fissionable plutonium, uranium and their mix as MOX—their production regulated, overseen, and controlled, not by a popular forum, but by Japanese electrical companies seeking great profit accumulations and operating under the political shield of two sets of state authorities—one the Japanese government hierarchy of in-bred bureaucrats working with conglomerated financial oligarchies; the other the U.S. federal government made up of look-alike political parties joined at the hip, lobbying for special interests controlling Wall Street and San Francisco, Los Angeles and Houston, Atlanta and Chicago.

These focal points of State authority and conglomerated wealth have made the nuclear output of vast amounts of energy the centerpiece of industrial and commercial operations in Japan, so much so that the risk of regional, even national, annihilation is rising. During overlapping natural disasters, it is conceivable that the meltdown of Nippon's fifty-four nuclear plants (along with the Rokkasho plutonium reprocessing facilities) might well ignite a chain reaction bringing on a cascade of lifelessness of the Japanese as a people.

Short of this final wave of death delivered, the nuclides released from the meltdown of even a few nuclear power plants holds the capacity to poison the regional atmosphere, groundwater, and soils on which millions of people and their domesticated crops and animals depend for life sustenance.

Supply lines of food and water to sustain prefectures also could be cut. Production with its essential workers would likely be disrupted for long stretches of time, ripping the social fabric. So enterprises unable to produce for markets would be undermined; and workers unable to earn salaries and wages to purchase their needs would be impoverished—all the while the flush of rarified upper classes and state authorities, who put nuclear power on the nation's agenda, would escape with their lives and wealth.

Lesser industries and smaller workshops, also lacking nuclear-generated power sources, would languish as the traditional shock absorbers of crises, unable to sell their goods to one another and the idled factory networks.

Their workers, too, would have neither jobs nor pay to buy essentials, sending the market economy into a tailspin. Whole communities might well perish economically and socially; and filial and emotional ties likely would be completely undermined, as relatives and once-neighbors became an uprooted, almost nameless, population on the march to other prefectures and cities in search of safer and sustainable conditions.

Industrial and urban centers might also fearfully and reluctantly receive the exodus of the desperate, now seen as tainted by nuclear poisons and sterility, whose poverty required special medical care and state aid and allotments from strained, already frail budgetary resources. The army of newcomers would become, then, not simply unwanted, but also an underclass of unemployed, untouchables perhaps, despised and hated, along with existing communities of Burakumin outcasts and isolated Ainu people.

Japan would be diminished as a great productive nation—not because of the still-vibrant, gentle nature of its people, but because the State-sustained ruling interests, lording over and viewing all of society in terms of accumulating wealth, refused to pursue the well-being of the nation's millions and the sustainability of the resources on which they depended for survival.

NOTE

1. Arnie Gundersen, "This Could Become Chernobyl on Steroids": Nuclear Engineer Arnie Gundersen on Japan's Growing Nuclear Crisis, March 15, 2011, http://www.democracynow.org/2011/3/15/this_could_become_chernobyl_on_steroids.

Chapter Two

Trial by Sea, Trial by Fire

At 2:46 p.m. on March 11, 2011, a huge rupture of an ocean subduction zone area spanning 400 by 200 kilometers produced a powerful 9.0 magnitude earthquake, triggering a series of tsunamis with 15-meter-high waves that crashed into the northeastern region of Japan known as Tohoku and reached as far as the U.S. west coast.[1] The Japan Meteorological Agency (JMA) and the Headquarters for Earthquake Research Promotion (HERP) estimated that the hypocenter of the earthquake was some 130 km off the coast of Miyagi Prefecture to the area off Iwate Prefecture in the north, and the area off Fukushima Prefecture and Ibaragi Prefecture in the south, at a depth of 24 km.[2]

Both tsunamis and earthquake wrecked nuclear power plants lining the Tohoku shores, igniting a chain of potentially cataclysmic explosions at the Fukushima Daiichi Nuclear Power Plant. Forty-five minutes after the initial earthquake, rising seawater at the entire Fukushima Daiichi plant disabled the entire seawater pump system, including eleven of the twelve emergency diesel generators, except the one connected to Unit 6.[3] The loss of nearly all power at the six reactors at the Fukushima Daiichi plant undermined all safety measures and nuclear core temperatures spiked.

For the next three days, the three reactors operating when the earthquake struck consecutively lost core cooling capabilities. A great mass of radioactivity from the damaged fuel escaped into the environment. As the cooling water boiled, pressure inside the reactor rose, creating leaks allowing more radiation to escape. Workers, attempting to reduce internal pressures, increasing radiaton release, vented the containments. Within four days of the quake explosions in the reactor buildings of units 1, 3, and 4 spewed out more radiation. The overheated reactors had melted their fuel, releasing hydrogen the buildup of which led to explosions.[4]

In addition to six reactors, the Fukushima Daiichi plant housed seven spent fuel pools. These no longer provided fission but still heated up enough to drop their coolant levels. The spent fuel rods never were exposed but the dirty pools became and remained uncovered, subject to dispersal by wind, rain, and leakage.

A WITHERING PICTURE OF DESTRUCTION

In Japan one observes development of human entropy. A civilization is being sundered, and a genetic cascade has already begun. Radiation will not destroy the Japanese and others it poisons; rather our species' hubris will be the agent of ruin. Our intellectual pride in our abilities to create Strangelove machines, not to mention fabulous short-term profit, has generated our cooperation with the forces of decay. The best and brightest—and most voracious—have wedded themselves to a Sisyphean task whose boulder is more lethal than his: it may not only roll back over him, but destroy all its shapers, users, funders, and perhaps life itself.

Modern times offer examples of such ecocidal dystopias. The first tsunami of death and disease related to nuclear use entered history with the blasts of American bombs at Hiroshima and Nagasaki in 1945. The Chernobyl nuclear complex in Ukraine exploded in 1986. Though some radioactive elements of its nuclear fuel quickly dissipated, others remained. The released cesium's original half-life would last thirty years and strontium's twenty-nine years.

Scientists today estimate that it takes ten to thirteen half-lives before biological life and economic activity can return an area near the size of Switzerland. The contaminated area—designated by Ukraine's parliament as 15,000 square miles—will be affected for over three centuries. At the same time, tourist business through the wasteland is thriving (even joked about as in "Visit Sunny Chernobyl"), some "returnees" legally or otherwise have taken up residences there, and Ukraine plans to continue nuclear research in the area!

In Japan, even should the best-case scenario be fulfilled, the Fukushima site has become a radioactive dump. Fukushima is now a massive storage facility for spent nuclear fuel rods, perhaps the most deadly and volatile form of nuclear waste. Cleanup errors, continuing leakages, and the dangers of what is something of a spread-out atomic device, and geological vulnerability of the former plant all add to possible slow onsets or explosions of entropy.

INTERLOCK: RING OF FIRE, RING OF LIFE

Fukushima may dwarf even the catastrophes of Hiroshima and Nagasaki. The Unit 1 (Daiichi) explosion came from the sea, from one tectonic plate sliding over another piece of deep Earth, and driving the tsunami tidal wave that destroyed its electrical cooling system. The heated-up plant shot radioactive gas and solids into the air and seeped poisons, deadly fissionable nuclides, into soil and groundwater.

The state authorities should have known—or cared. A 1995 disastrous 6.9-magnitude Kobe quake in southern Japan did nudge the government to require power plants to improve nuclear safeguards and construction standards, based on historic seismic activity in each region. However, the agency that had spent several years evaluating the Fukushima Daiichi nuclear plant had neglected to employ a tsusnami expert. It gave its assurance of safety even after its own experts pointed out that a tsunami could jeopardize its reactors.

For example, two years before the Fukushima destruction, expert seismologist Yukinobu Okamura warned of a potential debilitating tsunami. He spoke out at one of several meetings held by the Nuclear and Industrial Safety Agency to evaluate the readiness of Daiichi and other nuclear power plants to resist natural disasters. An executive from the Tokyo Electric Power Company (TEPCO) opposed Okamura, given that the power company and the government posited earthquakes as a greater threat. Okamura had on his side historical evidence that in 869 AD, a massive sea-quake near Sendai in northeastern Japan sent a tsunami wave more than two miles inland. The story had been treated as legend although twenty-first-century geologists inspecting earth strata and sediment began to substantiate the old story. But Sendai and Okamura were sidelined.[5]

Fukushima Daiichi's safety measures were given the seal of approval and thus left open to what unfolded when the 9.0-magnitude earthquake struck off Japan's northeast coast, killing generators and the cooling system. "The diesels were in a very low area," said Ken Brockman, former director of nuclear installation safety at the U.N.'s International Atomic Energy Agency. "That would make them very susceptible to a tsunami or even an internal flooding event."[6] Various experts confirm that Daiichi failed to withstand the earthquake. The ensuing tsunami, with waves more than thirty-two feet high, also knocked out the facility's critical backup power supply and triggered a nuclear emergency, resulting in widespread releases of radiation.[7]

The recent exposure of U.S. State Department documents by Wikileaks also show that the U.S. knew about the vulnerability of the Fukushima plant. U.S. cables describe serious concerns about safety, national security, and potential corporate liability for proprietary technologies, including nuclear reactors and electric power plants built in Japan. One U.S. diplomatic cable dated December 17, 2008, classified "sensitive," indicated that an international atomic energy official reported ". . . guides for seismic safety [in Japan] have only been revised three times in the last 35 years . . . [substantiating that] recent earthquakes in some cases have exceeded the design basis for some nuclear plants, and that this is a serious problem" (US Cable, 183350).

Another cable disclosed how the Japanese government opposed a court order to shut down a nuclear power plant in Shiga Prefecture, which was only "built to withstand a magnitude 6.5 earthquake" (U.S. Cable, 58190). The Shiga nuclear power plant was commissioned in 1993 using the General Electric BWR (Boiling Water Reactor) design, the type used for construction of the Fukushima plant. In 1976, three GE nuclear scientists blew the whistle on safety problems of the Boiling Water design for Fukushima (U.S. Cable, 149131, Section 29). The U.S. cable noted the structural flaw of the Shiga nuclear power plant in Shiga, a twin of the Fukushima plant. There should have been no surprise when the March 11 earthquake, eighty-one miles offshore from Japan, with a tremor near 9.0 magnitude ravaged the nuclear power plants in Fukushima.

GOVERNMENT CALIBRATIONS

The disaster shows the government's error in rating one natural disaster more of concern than another. It cast scrutiny and suspicion on reviews that amounted to little more than seconding "enlightened self-interest" codes of the Japanese Nuclear and Industrial Safety Agency (now renamed the Nuclear Regulatory Agency) and their designer, Tokyo Electric.[8] Backroom bargaining over risk was in progress the very moment the March 11th tsunami struck. By the next week, when the crisis was spinning out of control, the Japanese had yet to learn of the extent of radiation danger soon to rip apart Japan's social and psychological fabric. By then the Daiichi nuclear power station had been hit by the explosions damaging its cooling functions. It had sent radiation wafting into Tokyo, more than 130 miles away. Then a second reactor unit ruptured, spurting toxic steam. Tokyo Electric took away 750 workers, leaving a crew of 50 struggling to lower temperatures in three reactors, and an overheated plutonium pond burning and releasing fissionable nuclides.[9]

This crew of fifty was doomed. Invisible nuclear bullets passed through their bodies, iodine-131 discharges destroyed their thyroid glands, cesium tore at their lungs. Their fate and end went largely unreported, but word of mouth spread the news, especially when the government took the baby step of setting up a twelve-mile-radius no-go zone around the smoldering ruin.

Scientists warned that in case of meltdown, twelve miles would be too little for the boundary of an exclusion area, but the speed of what might amount to an atomic bomb blast would prevent efforts to avoid it and its effects. On March 19th, the Japanese still dickered between moving people farther away and the possible unprecedented—and unplanned-for—catastrophe. The lives of tens of thousands of people were on the gameboard.

Moreover, economic factors took a hand in the game. In the past, trying to offset declining northeastern coastal towns lacking job opportunities and losing population, the government had promoted nuclear power plants among them. These plants were bringing in $20 million or more in revenue. One of them, in Fukushima Prefecture was now in flames, enveloped in fumes, and spreading radiation through all the four elements. Economic depression led to the raw mood itself: many in the Northeast vowed to depart, others went blank when officals like Prime Minister Abe promised reconstruction and government help with the problem he saw as primary in the area, psychological surrender (Address by Prime Minister Shinzo Abe at the Ceremony to Commemorate the Third Anniversary of the Great East Japan Earthquake).

TRIPLE THREAT FORCE: EARTHQUAKE, TSUNAMI, NUCLEAR BREACH

Japan lies within the so-called "Ring of Fire"—earthquake and volcanic zones in a horseshoe shape around the Pacific. Here occur some 90 percent of the world's earthquakes and volcanic eruptions. Records of undersea quakes go back thousands of years, few as grim as the one that triggered a tsunami in the Indian Ocean on December 26, 2004, that killed an estimated 230,000 people in twelve countries. A magnitude 8.8 temblor shook central Chile in February 2010, generated a tsunami that killed at least 524 people.[10] A NASA chart representing eruptions and earthquakes with dots shows Japan blackened completely, according to NASA's charting.

Submarine earthquakes or volcanic eruptions can cause tsunamis, as could underwater nuclear explosions. But the sea-quakes are the best-known dangers since they bring on both shaking and flooding, as well as the scraping of the landscape as the heavy volume of waves drags itself back into the ocean. Plate tectonics, the geology of the earth's crust, explains the process. One plate of the earth's crust pushes above another, is resisted, and then builds mass, pushing the other plate down and itself collapses. The energy of that great splash generates, not a tide but a surge of water, sometimes called "a wave train," whose peaks are smaller than surf but capable of speeding landward at a speed of up to 450 miles per hour (750 kmph). The tsunami first appear to be merely an abnormal tide, but its chewing the landscape and perhaps sucking sound give what has often been a warning too late. The eating away of obstacles to its drive does not lessen the energy of the charge from the sea—given or making a favorable coastal platform, a tsunami may wreak havoc in areas for miles inland, both in its coming and in its going.

The power companies, the government, someone should have heeded an early warning for Fukushima Prefect on March 9, when several quakes, one of magnitude 7.3, hit the region, but there was little damage or concern about aftershocks or new danger. Two days later, the northeast coast shook under the pounding of a 9.0 quake—the largest to hit Japan since record-keeping began in the 1800s—clearly one of the biggest ever recorded in the world. Soon Japanese TV, even a certan pro-nuclear network, became a kaleidoscope of apocalyptic images of powerful, debris-filled waves, uncontrolled fires, a ship caught in a massive whirlpool. Large fishing boats and other vessels tumbled ashore on great waves, slamming against and collapsing overpasses, upending roads, snapping power and telephone lines. Power-packed surf and irresistible undertows rattled like baby toys shipping fleets moving or anchored. Roads cracked and cratered, throwing vehicles into the surge where they bobbed about as if pool playthings before sinking. The tsunami roared over embankments, washing inland anything in its path, then powerfully reversing to suck homes, vehicles, and tons of debris out to sea. Few survived the grasp of the stormy jaws, few were the victims from them recovered. The tsunami's path inland was marked by fire set off and spread from burst gas lines and downed power connections.

Some drivers attempted to flee, but potential evacuation routes, unmarked and unregulated, trapped them in jams stretching many kilometers. Sendai airport was covered in slurry and debris that stopped everything on wheels; Tokyo's Narita airport was closed indefinitely. In the worst-hit coastal areas roads collapsed and larger highways and overpasses buckled. Telephone lines stretched, then frayed, and communications all across the country ceased. Some 10 million daily commuters stood askance, waiting in northeastern Japan and in Tokyo for trains that never arrived as the authorities suspended service.[11]

As of 2013, the government reported disaster deaths of 15,880 (15,814 in the three northeast prefectures of Fukushima, Miyagi, and Iwate); missing and presumed dead: 2,994 (2,690 from the northeast); injured: 6,135 (4,534); number of dwellings completely destroyed: 128,931 (124,900); dwellings partially destroyed: 269,045 (231,571).[12]

U.S. RESPONSES TO THE FUKUSHIMA DISASTER

General Electric designed the six reactors at the Fukushima nuclear power plant and supervised their start-up in 1971–1979.[13] The disaster there accordingly triggered quick denials and defense from GE, seconded by the U.S. government. GE refused any reponsibility for the blowup. GE's first comment was "The GWR Mark 1 reactor is the industry's workhorse with a proven track record of safety and reliability for more than 40 years. Today, there are 32 BWR Mark 1 reactors operating as designed worldwide. There has never been a breach of a Mark 1 containment system."[14] GE said the ruined reactor "performed as designed," even after the shock of a 9.0 earthquake.[15] GE emailed the full twenty-one-page justification paper to media people after unflattering reports on the history of the Mark 1 reactor used in Fukushima. Nonetheless, questions have persisted for decdes about the ability of the Mark 1 to handle the immense pressures resulting from the loss of cooling power in such reactors.[16]

Dale G. Bridenbaugh, one of three GE nuclear scientists and whistleblowers who resigned from their positions in 1976, said on March 15 that "the problems we identified in 1975 were that, in doing the design of the containment, they did not take into account the dynamic loads that could be experienced with a loss of coolant. . . . The impact loads the containment would receive by this very rapid release of energy could tear the containment apart and create uncontrolled release."[17] His fear of a devastating accident because of flaws in design and the danger of the Mark 1 nuclear reactor led to his increased anxiety and resignation.

The GE system, originally meant to propel the first generation of U.S. nuclear submarines, situates cooling pools for spent fuel rods directly over the top of live nuclear reactors. Any disruption of a reactor was thus likely to interrupt the mechanism for storing nuclear waste. The U.S. Atomic Energy Commission in 1971 expressed worries, too, about design flaws in GE's Mark 1 nuclear reactor.[18] Top AEC safety official Joseph Hendrie's memo of September 25, 1972, recognized that the GE reactor design could, in an accident, enable a core meltdown, but he refused to say because such a risk, when known, "could well mean the end of nuclear power."[19]

America's largest energy company, NRG Energy, got moving after it learned of the doubting. Before the disaster, it had wrangled a $4 billion loan guarantee from the Obama administration for its South Texas Project about ninety miles southwest of Houston. The two nuclear reactors there were to be built and operated by Tokyo Electric and local partners. NRG and the Japanese further contracted to build other nuclear plants in Texas.[20]

In August 2006 NRG asked Tokyo Electric to provide technical consulting services and invest in the project, a request accepted and acted on in March 2007 after the Japanese announcement that the decision was "based on the stable long-term earnings expected from the project, as well as . . . contributing internationally towards the realization of a low carbon society and fostering human resources related to the nuclear power business."[21] But a month after the Fukushima cataclysm NRG, citing the disaster, shredded the agreement.[22]

The U.S. military and embassy also snapped to, expanding the no-go zone around the devastation to fifty miles. The embassy ordered U.S. citizens in that zone to evacuate or stay indoors.[23] Military medics advised U.S. rescue mission crews flying within eighty miles of the nuclear complex to take potassium iodide tablets to combat radiation effects, and the U.S. embassy began issuing such pills to U.S. civilians.[24],[25]

These, fearful, packed escape flights leaving the country, the first of these out of Yokota Air Base on March 19.[26] Nearly all 240 seats were filled and the plane was able to place all residents who wanted to leave on the flight. Misawa Air Base in northern Japan was another major exit point.[27] By March 25, two weeks after the earthquake, more than seven thousand U.S. family members from five military bases left for the U.S. under the military's "voluntary departure" plan.[28]

On March 12, the U.S. military in Japan also instituted "Operation Tomodachi" to provide support in disaster relief, which lasted until May 4, 2011. The operation involved 24,000 U.S. service members, 189 aircraft, and 24 naval ships at a total cost of $90 million.[29] U.S. forces helped rescue about twenty thousand people in the first week after the quake, and worked to restore transportation facilities such as the Sendai International Airport.[30]

CONCLUSIONS

Japan's response to the nuclear crisis after the March 11 tsunami and earthquake was confused and problematic. A 507-page interim government report, compiled by interviewing over four hundred people and published in December 2011, revealed false assumptions that an emergency cooling system was working, adding to a delay in disclosing dangerous radiation leaks.[31] The report also showed that authorities had grossly misjudged tsunami risks, and it criticized the use of the term, "*Soteigai*," meaning "outside our imagination," implying that authorities shirked guilt about their neglect.[32]

The U.S. acted more decisively and quickly, ordering evacuation of its citizens from the disaster area, and assisting families of service members to get out of Japan. General Electric washed its hands of any connection to the spectacular calamity and dragged across its trail of cuplability the red herring of the trouble-free operation of its products elsewhere. The CIA, and former Class-A war criminal Matsutaro Shoriki rolled with the punch and made sometimes separate plans to nurture nukes and nationalism; the disaster, along with opening doors in London, gave Shoriki more bargaining power, just as the reddening of the East and the occasional missile launch from

North Korea, so inept that some wondered if Kim Jong-il was using slingshots as launch devices, amplified U.S. lust for a strong Asian ally.

In 2014, nearly three years after the Fukushima disaster, Fukushimites brought suits against GE and two Japanese corporations, Hitachi and Toshiba, that helped build the Fukushima nuclear plant.[33] The 1,400 plaintiffs, many of whom resided near the nuclear plant, said that the corporations should be financially liable for damages caused by the reactors' meltdown.[34] The plaintiffs stressed that their first priority was not the economic compensation, but to raise awareness about the problem with the power plant manufacturers' immunity and the danger of nuclear energy.[35]

The health of disaster area residents obviously was fragile. A September 30, 2013, screening of Fukushima residents eighteen or younger when the plant blew had found twenty-four confirmed and thirty-two suspected cases of thyroid cancer.[36] In 2013, the World Health Organization estimated a 70 percent higher than average risk of radiation-exposed female infants developing thyroid cancer, a 7 percent higher risk of leukemia for exposed infant males, and a 6 percent higher than average risk of breast cancer in females exposed as infants.[37] But Prime Minister Abe insisted that the main problem was psychological.

Also Dr. Shunichi Yamashita, head of the Fukushima Health Survey following the accident, asserted that the radiation leak was not serious. He persuaded the officials of Fukushima Medical University there was absolutely no need to pass out potassium iodide pills because the spread and effect of radioactivity from the accident would be miniscule compared to Chernobyl.[38] Yamashita was a professor of radiation medicine at Nagasaki University, then an adviser to Fukushima Prefecture on radiation health risk, mainly fear, control: "People are suffering, not only because of the earthquake and the tsunami, but also from severe radiation anxiety, real radiophobia. . . . [In Chernobyl] life expectancy of the evacuees dropped from 65 to 58 years—not because of cancer, but from depression, alcoholism, and suicide. Radiation is hard; the stress is huge."[39]

The UN and government leaders chimed in. The 2012 World Health Organization said a little fission never hurt anyone and even a dose of 50mSv of radiation posed "a fairly minimal [health] risk."[40] Chief Cabinet Secretary Yukio Edano repeatedly stated on Japaanese TV that there was no "immediate health risk." from the level of radiation detected some twenty kilometers from the Daiichi plant.[41] The pro-nuke claque drowned out a few objectors such as Hisako Sakiyama, of the National Fukushima Nuclear Accident Investigation Committee, who charged that the developing nuclear safety myth was based on misinformed consent.[42]

The next chapter will examine the impact of nuclear fallout on those exposed to it, including Fukushimans and U.S. servicemen in Operation Tomodachi just after the March 11th earthquake.

NOTES

1. Larry Greenemeler, "How Does an Earthquake Trigger Tsunamis Thousands of Kilometers Away?" *Scientific American*, Mar. 11, 2011.
2. Report of Japanese Government to the IAEA Ministerial Conference on Nuclear Safety (hereinafter IAEA Report), June 2011, at III-1 & 2, available at http://japan.kantei.go.jp/kan/topics/201106/pdf/chapter_iii-1.pdf.
3. James M. Acton and Mark Hibbs, "Why Fukushima Was Preventable," *Carnegie Papers*, March 2012, at 5.
4. Id., at 6.
5. David Nakumura, "Japanese Nuclear Plant's Safety Analysts Brushed Off Risk of Tsunami," *The Washington Post*, Mar. 24, 2011.
6. Richard Lewis, "Even the Best-Laid Plants . . .," *Citizenvox*, Mar. 24, 2011.
7. IAEA Report, at 28-35.
8. Nakumura, supra note 1.
9. "No Happy Ending," *Before It's News*, Mar. 16, 2011.
10. "Survivors Commemorate Asian Tsunami," *IntellAsia*, Dec. 27, 2012.
11. Stephen Harer, "At a Two Year Anniversary, the Tohoku Earthquake/Tsunami and Fukushima Nuclear Disaster Continue to Exact a Toll," *Forbes*, March 11, 2013.
12. Id.
13. Id.
14. Bill Dedman, "General Electric–Designed Reactors in Fukushima Have 23 Sisters in U.S.," *NBC News*, Mar. 12, 2011.
15. Matthew Most, "Fukushima: Mark 1 Nuclear Reactor Design Caused GE Scientist to Quit in Protest," *ABC News*, Mar. 15, 2011.
16. Matthew Most, "General Electric Defends Nuclear Plant Design," *ABC News*, Mar. 21, 2011.
17. Id.
18. "Law to Relieve GE of Liability for Reactors at Fukushima," Vernon Radiation Safety, (last visited on Feb. 6, 2014), available at http://vernonradiationsafety.wordpress.com/2013/12/05/law-to-relieve-ge-of-liability-for-reactors-at-fukushima/.
19. Id.
20. "TEPCO to Invest in South Texas Project Expansion (STP 3&4): First Japanese Utility to Invest in the Overseas Nuclear Power Project, *TEPCO*, May 10, 2010, available at http://www.tepco.co.jp/en/press/corp-com/release/10051001-e.html.
21. Id.
22. Matthew Wald, "NRG Abandons Project for 2 Reactors in Texas," *New York Times*, April 19, 2011.
23. Chris Caroll, "Pentagon Preparing for a Nuclear Worst-Case Scenario at Fukushima," *Stars & Stripes*, Mar. 16, 2011.
24. Id.
25. "U.S. Embassy Stops Distributing Iodide Pills to Citizens in Japan," *Stars & Stripes*, Apr. 18, 2011.
26. T. D. Flac, Travis Tritten, and Erik Slavin, "First Military Evacuation Flight Leaves Japan," *Stars & Stripes*, Mar. 19, 2011.
27. Id.
28. Seth Robson, "Streets of Tokyo, Yokota AB Remain Quiet Two Weeks After Quake," *Stars & Stripes*, Mar. 25, 2011
29. Eric Johnston, "Operation Tomodachi a Huge Success, But Was It a One-Off?" *Japan Times*, Mar. 3, 2012.
30. Id.
31. Susie Madrak, Report: Japan's Response to Fukushima Disaster Was a Mess, *Crooksandliars.com*, Dec. 27, 2011.
32. Id.
33. "Hundreds Sue Toshiba, GE, Hitachi Over Fukushima Nuclear Disaster," *The Voice of Russia*, Jan 31, 2014.

34. Id.
35. Id.
36. Mizuho Aoki, "Thyroid Cancers Up in Fukushima," *Japan Times*, Nov. 11, 2013.
37. "Fukushima-Area Cancer Risks Are Higher Than Normal After Japan Nuclear Disaster, WHO Reports," *Huffington Post*, Feb. 6, 2014.
38. J. D. Heyes, "Radiation Dunce Dr. Shinichi Yamashita Finally Admits Giving Bad Info About Fukushima Fallout," *Natural News*, Nov. 22, 2013.
39. Cordula Meyer, "Studying the Fukushima Aftermath: 'People are Suffering from Radiophobia,'" *Spiegel International*, Aug. 19, 2011.
40. Geoffrey Brumfiel, "World Health Organization Weighs in on Fukushima," *Newsblog*, May 23, 2013.
41. "NHK World Edano No Health Risk 20km Away From Plant," *Allvoices*, Mar. 16, 2011.
42. "Shinpojumu Fukushima Genpatsu de Naniga Okitaka – Anzen Shinwa no Hokai Hokoku [Symposium: What Truly Happened at Fukushima Nuclear Plant: Deconstruction of Safety Myth]," 15 *Simin Tsushin* 1, 10 (2012), avalable at http://archives.shiminkagaku.org/archives/csijnewsletter_015_nagasoe_121225.pdf.

Chapter Three

Ingenious Invisible Invaders

Immediately after the Mark 1 reactor explosion in Fukushima, Takayuki Matsuda, Health Ministry spokesman, advised a village near the crippled nuclear plant not to drink its tap water because it showed elevated levels of radioactive iodine. He said it was only a precaution, the increase being only one twenty-sixth of the level of a chest X-ray and thus harmless. His assurance was an early and much-repeated gambit in creating and spreading what would come to be known as "the safety myth." The myth pioneers, even though their way was largely clear, occasionally had to circle their wagons against incursions of fact. An early instance was the finding by health officials who imprudently let it be known that traces of radiation had also been found in the water of Tokyo and other areas and in spinach, milk, chrysanthemum greens, and canola.[1]

Tokyo Electric, too, minimized dangers from and denied any corporate responsibility for the failure of its Fukushima Daiichi plant and armed itself for a legal scrap possibly influencing the future of nuclear power worldwide. The utility first insisted that the magnitude 9 earthquake on 11 March was beyond anyone's expectations, impossible to guard against. However, testimony from designers and regulators, as well as leaked documents, began to strengthen the view that the company had scrimped on costs by cutting corners in building and as a result increasing the plant's vulnerability.[2]

Consciousness of mass disasters has a short shelf life, especially among those who inflict them. Who in the U.S. gives a thought to the firebombing of Tokyo by 330 B-29s the night of March 9, 1945, that killed over 100,000 residents and went down in the pages of infamy as the single most destructive bombing raid in history,[3] far more than the more publicized destruction of Dresden with some 3,000 tons of bombs in February, leaving a firestorm and an estimated 50,000 dead, itself largely forgotten in the West except by fans of Kurt Vonnegut's best-seller *Slaughterhouse-Five* or the film, *Dr. Strangelove, or How I Stopped Worrying and Loved the Bomb,* with a world-ender directed against a Russian target. Tokyo was almost forgotten, too, when a nuclear uranium bomb fell on Hiroshima and a plutonium fission bomb hit Nagasaki. These nuclear weapons of mass destruction were tools of experiments subjecting an entire urban populace to blasts, radiation, and chaos; the experiments put into play by the collaboration of American nuclear scientists, generals, and President Truman's cabinet, privy to the almost-complete Japanese negotiations for surrender.[4] Is Hiroshima Day on anyone's calendar except the Japanese? Do Western nations know more than Justin Bieber about Asian issues?

Of course, the Strangelovian Trojan horse "Atoms for Peace" and other initiatives arose primarily from U.S. desire for a bulwark against Mao and Stalin and profits from technology sales and service. There would be blowback, however. The U.S. spent tens of billions of dollars creating the nuclear industry at home and selling it overseas as safe, cheap, and limitless, but given other countries' equally determined desires for profit, prestige, and security, the monopoly seeped secrets and began to break up, with nuclear weapons now in the hands of major nations East and West, as well as Israel, Pakistan, North Korea, and probably other regimes in the Middle East, Africa, and South America. Prices fall as export beomes more competitive, Russia is tuning up nuclear capabilities in East Europe, and France, England, and China are developing nuclear energy and weapon wares for use at home and abroad.

Japan at mid-twentieth century was something of a captive audience for all things American, even baseball. It was seen as and became the best customer for U.S. nuclear sales but, given its past ambitions, required carefull pulls of the velvet leash of Atoms for Peace and the Atomic Energy Act, relying on the facts that Japan possessed no uranium deposits of its own and at the time had only limited capacity to enrich the mineral for use as nuclear fuel. The U.S. could regulate how U.S.-owned plutonium was used or processed by others. Japan became America's largest single foreign customer for enriched uranium. By 1988, the U.S. accounted for about one-half of Japan's needs, producing one-quarter of its electrical power. But some of the friends of the atom inside Japan, corporate nationalists and imperialists, began to wriggle free, by playing former allies off against each other, courting the English atomic developers, importing a French pilot reprocessing plant and beginning a nuclear fuel–reprocessing plant and spent-fuel facility at Rokkasho Village—steps toward a fully independent nuclear industry, civilian and military.[5]

Scientists with nuclear programs in the fifties and sixties knew that such technology involved jeopardy, but put such concerns far behind nurturing the bugeoning industry. Risks and critics were muted or went unaired. No public debate on risks versus benefits of nuclear-generated power ever hindered the hyperbole, misstatements, or misdirection of interests that might have questioned the nuclear hucksterism of Atoms for Peace and safety myth purveyors. Less than a decade after Hiroshima, however, scientists knew that radioactive elements, isotopes, generated by an atomic explosion went into the atmosphere in tiny fallout particles, contaminating food, water and air, absorbed by the human body. Their concentration in various body organs, though, remained a mystery. In the body, they apparently behaved just like their natural unradiating counterparts. But their rays were actually invisible bullets wounding the atomic structure of these organs. A biologist might equate them to malignant viruses with the power to linger, weaken, and kill.

MARKETING (AND PAPERING OVER) MENACES

Just like bone and tooth sustain calcium, strontium-90 settles in these structures with the more malignant effect of punching holes in their molecules. In the iodine-producing thyroid gland vital to the growth and functioning of the human body, radioactive iodine-131 mimics regular iodine—except it ends thyroid function. Researchers in the 1950s found it concentrated in the milk of cows grazing in pastures contaminated by fallout. People drank the milk, the iodine built up and concentrated in their thyroid glands, and became particularly lethal because of the naturally small size of such glands.

An adult thyroid could absorb one hundred times an external dosage. But since perinatal glands are smaller, concentration density in the glands of unborn children and infants can reach ten times that. That is, environmental contaminants could be concentrated by a factor of ten in adults and one thousand in children[6]—and cancer can be triggered by a single particle of radiation scoring a hit on a crucial chromosomal DNA molecule. Even non-lethal wounds to the thyroid may induce infertility, microcephaly, depression and fatigue, mental confusion, hair loss, body bloating or anorexia, and a slow onset of metastatic cancer. Radiation damage to the DNA of any body cell could produce cancer because DNA contains the genetic code controlling cell functions and reproduction. Harmed by radiation, DNA can lose control and permit a cell to begin the unlimited growth of cancer. Massive or long-term absorbed radiation from fallout means cumulative genetic, possibly heritable, damage.[7]

Scientists such as Dr. John Goffman, former Director of the Department of Energy's Lawrence Livermore National Laboratory, maintain that there is no safe dose of radiation and its effects are cumulative.[8] Even his well-funded opponents concede that low-dose exposure is more harmful than previously thought. Organisms experiencing a high degree of cell division, as in fetuses and infants, are at greatest risk, not only of cancer, but many immune system problems and precursors of Down's syndrome, hemophilia, and muscular dystrophy.[9] However, such warnings about radiation have been muted by promises of its promise.

In weighing health effects of two or three Japanese nuclear plant meltdowns, one might consider the 1961 Russian detonation of a 100-megaton hydrogen test bomb over Siberia: following the test of the superradiant weapon, fissioning iodine showed up in milk throughout the northern hemisphere. In 1973 Professor Ernest J. Sternglass, Director of Radiological Physics in the University of Pittsburgh School of Public Health, explained: "Nuclear power reactors, atomic gas-mining explosions, and other forms of nuclear engineering, all normally release low levels of radioactivity and, in the event of an accident, they entail the risk of much worse. . . . those individuals who had been in positions of responsibility would have a terrible guilt to bear for the damage already done."[10] This guilt, however, has been callously denied in Japan and the U.S.

For example, less than two weeks after the Fukushima explosions, Evan Douple, associate chief of research at the Radiation Effects Research Foundation, avowed that "on the basis of our current estimates, there shouldn't be measurable numbers of cancers. So you won't be able to count them, ever. But once the dose estimates are put together and extrapolated, you should be able to make a crude estimate of the health effects, based on the Radiation Effects Research Foundation data. And I think that estimate will surprise a lot of people [because] they're so low."[11] Prime Minister Abe said the main Fukushima meltdown damage was psychological: you could read or hear as much in Japanese media. On U.S. TV, you could see baseball games or worries about the possible shutdown of the federal government and the actual shutdown of the National Football League.

Eventually, though, many Western industrial countries cooled in their avidity for the atom. In Year 1 A.D. (After Disaster), Germany shut down eight reactors and pledged to close the rest by 2022.[12] Polls in France, where 80 percent of all energy came from nuclear power, showed 83 percent of the public against new nuclear plants for rising energy demands. A Canadian poll found a majority opposing nuclear power energy, and the Province of British Columbia declared itself a nuclear-free zone.[13] Switzerland and Belgium approved phasing out all nuclear power before mid-century.[14]

Given a situation similar to the rise in opposition to smoking, nuclear corporations, as did the tobacco industry, increased marketing efforts in developing economies—particularly in Asia and Eastern Europe. No new nuclear reactors have been ordered in the U.S. since 1973, but the technology regarded as prolematic or of marginal profitability is promoted abroad by U.S., Canadian, European, and now Asian corporations.[15] Japan planned to boost nuclear exports even after the disaster and agreed in October 2011 to export Japan's nuclear technology to Vietnam.[16] Japan also signed an energy act with the Turkish government in January 2014 as a first step to exporting nuclear technology there.[17] The International Atomic Energy Agency (IAEA) meanwhile was created to achieve the contradictory aims of promoting nuclear power and preventing the spread of nuclear weapons. Atoms for Peace, however, continued a tendency to support the deployment of military molecules.[18] Its director Yukio Amano rose through nuclear control and developent posts and was reported to have described himself as "solidly in the U.S. court on every key strategic decision . . ." (*The Guardian*, November 2010) and was accused in March 2012 by several former senior IAEA officials of pro-western bias, overreliance on unverified intelligence, and sidelining sceptics.

Yet the "Atoms for Peace" propaganda campaign rolls on, spreading the safety myth of nuclear energy and mulching the growth of a treacherous technology. Nuclear power has never been, nor will be, safe.[19] Even chronic low levels of exposure—like those that would occur during normal transportation and from living in an area contaminated by an accident—have serious if slow-onset effects. However, the nuclear promotion hoopla has apparently drowned out whistle-blowers. Even as Fukushima reactors continue to exude poisons, five new nuclear facilities are planned in Texas, and international marketing goes on. Some very forward-thinking coporate grantees even discusss shooting nuclear wastes into space. And, as of 2014, Chernobyl is still dangerous and a giant metal shield is only in the process of being placed over the damaged nuclear plants in Fukushima.

HUBRIS, HEALTH, AND IONIZING RADIATION

Nuclear technologies and "normal" as well as indirect radiation—eating, breathing, or contact with radiated matter or exposures from X-rays, sunlight, or tanning equipment—subject human tissues and organs to three types of atom particles that actively affect living matter. These are classified as alpha, beta, or gamma rays according their power to penetrate and how they affect human tissue.

All are dangerous with cumulative effects. Alpha particles are, relative to the others, the largest rays and most sluggish but are also greatly ionized, and thus prone to cause the loss of electrons in contacted atoms. They can devastate that zone of contact. Paper and skin repel them, but, inhaled, they adhere to lungs and continually irradiate the nearby area. Gamma rays are sparsely ionizing but of higher energy and penetrability.They are invasive, causing less damage per area than alpha particles, but affecting larger areas of tissue and blood. They pass right through the metal sides of the waste and transportation containers and into the human body.

Beta particles fall somewhere in between. They are stopped by metal but not by paper or skin and carry a fissionable power-pack greater than alpha particles. The radiation in contact-handled trans-uranic waste is mostly in the form of alpha particles, but also contains some gamma radiation because the trans-uranics decay into other elements like americium which emit gamma rays.[20] What it comes down to is elementary and frightening. Nuclear pollution, what counts biologically, is the sum of all the injuries over time from all the combined sources and events which release persistent poisons (radioactive or other) into the biosphere—so each contribution to the sum matters.[21]

In summary, normal nuclear production sends poisonous radioactivity into the air, soil, and water. Radioactive substances emit alpha and beta particles and gamma rays which can harm living cells. A high dose of radiation can lead to death within days or weeks, and low doses of radiation are now known to be much more damaging to health than previously thought.[22] Ongoing exposure to so-called low-level radiation can cause severe, enduring human health problems, both to those exposed and, through genetic mutation, to their descendants. The scientific community has reached a consensus that radiation exposure at any level is dangerous.[23]

THE SAFETY MYTH OF LOW-LEVEL RADIATION

Repetitious governmental misinformation and purposeful failures to protect the nation's children can lead to many hundreds of thousands, if not millions, of young people suffering with dysfunctional thyroid glands, stunted development, and cancers. Government and corporate media machines still propagate the myth that low-level radiation even in Fukushima posed and poses no harm to adults, children, and life in utero.

Harmlessness was a lie; the myth an untruth, found Toshiso Kosako, professor and radiation safety expert at the University of Tokyo. He wept and resigned his post as Prime Minister Naoto Kan's adviser to the Japanese Cabinet during a dramatic news conference. He accused the government of failure to protect the population from radiation. One of his most damaging charges was that the government permitted exposure levels for children living near the crippled nuclear plant to be set at international standards for nuclear power plant workers, a level far higher than standards set for the public.[24] "I cannot allow this as a scholar," said Professor Kosako, also blasting the government for what he said was a lack of transparency in releasing radiation levels around the Daiichi plant, and for allowing overly high exposure limits for workers who spent weeks struggling to keep the plant under control.[25] Kosako's abrupt resignation fed growing criticism of the government's handling of the crisis. More people suspected the government of minimizing the continuing dangers the plant posed and the amount of radioactive material released.

Disputing these criticisms before Parliament, Prime Minister Kan defended his government's handling of the nation's nuclear crisis, still uncertain what to do about the most contaminated land around the plant and the towns outside the evacuation zone reporting relatively high levels of radiation. It was a battle of protecting the central government's budgetary resources against protecting human life and environmental safety. For any decision about the levels of permissible radiation exposure could force expensive soil cleanups—inflating the already staggering estimates of the cost of rebuilding, $30 billion in a May 2011 estimate. Even widening the evacuation and no-go zone would be a big blow to densely populated Japan, as well as farms and fisheries in its northeast.

The disaster and attempts to silence or deny related information put a sharper edge on class divisions. Farmers and fishermen worried about more disclosures of radiation that would destroy consumers' faith in their goods and their livelihoods. Some of them worried about the effect of tainted products on others and took to heart calls from parents and ecology defenders for the government to lower permissible exposure limits. Another issue took form as concerned citizens realized that highly contaminated schools indicated even worse pollution in the neighborhoods around them.

Some fearful populations gave up waiting for government action or even pratical planning. One city in Fukushima, Kodiyama, thirty-five miles from the Daiichi plant, announced a municipal cleanup, the removal of topsoil at fifteen elementary schools where radiation levels had been detected in excess of the standard 3.8 microsieverts per hour.[26]

FURY AND MURDER

Just one day after the giant tsunami set off the continuing disaster at the Daiichi nuclear plant, thousands from nearby Namie were assembled for a hurried flight amid a gathering storm of anger as they learned of unpublished radiation forecasts which might have reduced their danger and harm. Blindly, given little or no guidance or aid from politicians or corporate oligarchs, most of the town groped its way north, trusting that usual winter north-to-south winds would blow radioactive emissions away from their exodus. As hydrogen explosions at the reactors spewed radiation into the air for three nights, city residents out in the country, but not far away from Namie, improvised shelters and moved on in patches. At stops not overrun with refugees, children played outside and families cooked and washed with—and drank—stream water. Unknown to the confused mob was that winds from Fukishima were moving not south but directly toward Namie and north in its Tsushima province. Government computerized forecast and detection systems showed them moving in a fog of predicted and verified contamination.

Such data was kept close to the chest of officials and in the safes of the coporations, reluctant to encounter criticism, to enlarge the ten-to-twelve-mile evacuation zone, and to admit their downplaying of the accident's reach and severity. However, Atoms for Peace had set off a very dirty bomb with effects difficult to deny.

The increasingly skeptical public doubted government's claims of concern and minimal danger. They recalled the response to a 1950s case, when a national scandal developed after people realized that government bureaucrats and corporate officials colluded to protect profits by trying to cover up a chemical factory's release of toxic mercury into Minamata Bay in the Kyushu Islands. As of March 2001, the chemical company had admitted to 2,265 seriously affected victims and compensated more than 10,000 of these.[27] The mercury leak caused nerve disease of those in the region, publicized by many wrenching photographs of stricken and deformed victims. Despite Minamata, the government sat on warnings about the probably greater spread of radiation from the Fukushima meltdowns, explosions, and leaks. It also failed to release information on the spread of radiation particles by a simulated computer system designed to forecast radiation dispersal.[28]

This warning and tracking system available to the government was called The System for Prediction of Environmental Emergency Dose Information (SPEEDI) and developed in the 1980s by the Nuclear Safety Division of the Ministry of Education, Culture, Sports, Science and Technology. Its data was generated by a network of nuclear detectors and had clearly shown that the Tsushima district would be susceptible to high levels of radiation from the Fukushima plant. Instead of using the emergency dose information and determining the relatively safest routes for evacuation, the government, after the disaster, simply told residents to run away and stay away.[29]

The Japanese, however, opened their files and projections to the U.S. forces a few days after the crisis. The Foreign Ministry disclosed more or less everything to the U.S. on March 14, but delayed for almost two weeks its selective release of similar information to the Japanese public.[30] Parents in the City of Koriyama in Fukushima said official statements were unrelible and brought suit against the national government in December 2011 in order to relocate their children to a safer area.[31] They also meant to challenge the nation's handling of evacuations and the public health crisis: it took two years for the case to come to a pre-trial hearing: the national court refused to hear the case.[32]

The parents, denied data and access, lacked proof of the cause-and-effect relation between the calamity and assorted illnesses and deaths, thus permitting Tokyo Electric and the government to evade resposibility for what happened in or near the Daiichi reactor. Many of the nuclear accident's health consequences, too, would become clear only years later. Meanwhile propaganda hindered discussion and the government raised twenty-fold the legal exposure limit to radiation, effectively allowing families to remain in contamination areas banned by the old legal standard. The limit was later scaled back to the old standard, one millisievert per year, but applied only to children while they were inside school buildings. Outside, with the nuclear strafing, they were on their own.

GOVERNMENT NON-DISCLOSURE OF INFORMATION ON RADIATION DANGERS

In late March 2011, approximately 45 percent of 1,080 children in three Fukushima communities tested positive for thyroid exposure to radiation, according to the government annoucement, which added the sop that the levels were too low to warrant further examination.[33] Experts in and outside Japan questioned the government's palliative, pointing out that most of those in Chernobyl who went on to suffer from thyroid cancer were children living near that plant when it blew. On March 15, Mr. Kan called Mr. Soramoto, former designer of nuclear plants for Toshiba, to ask for his help in managing the escalating crisis. Mr. Soramoto formed an impromptu advisory group including his former professor at the University of Tokyo, Toshiso Kosako, an expert on radiation measurement.[34]

Professor Kosako, who had studied the Soviet response to Chernobyl, said he was surprised at how little the prime minister's officers knew about available monitoring resources. He quickly advised Chief Cabinet Secretary Yukio Edano to use the enviromental emergency dose information and weather and topographical data to sharpen predictions of the spread of radioactive materials after their Fukushima plant rampage.

Indeed, The System for Prediction of Environmental Emergency Dose Information was producing maps and other data hourly the first hours after the earthquake and tsunami. But the Ministry of Education, Science and Technology had not shared the data with the prime minister's office because, said an internal memo, information was incomplete. The tsunami had knocked out sensors at the plant, and thus, without measurements of how much radiation was actually being released by the plant, it was impossible to predict how far the radioactive plume was stretching.[35] However, the internal memo of March 2012 revealed that estimates and forecasts seemed too serious and upsetting for release—the real reason to keep them quiet to disclose to the public.[36] The first internal memo went out on March 15, 2011, immediately after the third reactor explosion at the Daiichi plant.[37] The second came out a year later.

Tragically, for nuclear refugees from Tsushima and other residents who were entrapped in radiation-contaminated Fukushima and adjacent prefectures and regions, genetic damage caused by radiation is heritable—potentially affecting future generations, as well as all living species, animal of and plants in these areas. The future of radiation danger and damage was the last thing the government, the nuclear industry, and the public media wanted known.

SAILING INTO HELL

In January 2014, seventy-one sick U.S. Navy sailors sent to Operation Tomodachi, disaster relief, filed lawsuits against Tokyo Electric. They charged that its sluggish cleanup and control effort had caused their cancers and other radiation-linked illnesses.[38] They voted with their bodies for a glimpse of truth and corporate responsibility and for the U.S to fulfill its duty to protect its military. Tokyo Electric and then the White House reponded with delay, whitewash, and stonewalling. The U.S. had already, from 1951 to 1957, exposed some 400,000 of its soldiers and civilians to atomic tests with high levels of radiation exposure.

A year before Operation Tomodachi, Edward Snowden made such information public, as well as the U.S. National Security Agency's snooping on leaders such as German and Brazilian heads of state and even the pope.[39] NSA has an especially sensitive ear, or global surveillance program, at its Signal Intelligence station at the Misawa Air Force Base near Fukushima. This station houses[40] an array of nineteen parabolic antennas inside radomes that pick up data from American satellites and foreign relays using satellites in geosynchronous orbit. With all its surveillance technology and Japanese translators to monitor the unfolding Tokyo Electric catastrophe, as well as taps, satellites, and drones, in the U.S. the security agency quite probably frequently updated the Obama administration on such situations and their effects.

Nonetheless, to fix blame, or repair damage from mishandling of information and access to and use of classified information remains problematic. One could not show, even post-Sowden, that NSA passed findings to the U.S. government nor even pin down the sources and go-betweens with the data. Trails in the Obama office were unmarked or obliterated, too. An opaque fog of NIMI (not in my in-box) and WLIT ("We're Looking into That") muffled and obscured who, what, or when questions. In January 2014, Congress finally instructed the Defense Department to review health impacts on Navy first responders in Operation Tomodachi.[41] It has yet to ask critical questions about the extent and handling of NSA information on Fukushima.

U.S. officials advised by experts well knew of the danger from the Daiichi meltdown, but they largely kept quiet—and were not asked—about the Navy lawsuits, or solemn predictions like that of the University of Hawaii's International Pacific Research Center in 2013. It predicted that "within 5-6 years, the emissions [from Japan] will reach the North American coastline, with uncertain consequences for food safety and health of the local population. . . . After 10 years, the concentrations [of radiation will] become nearly homogenous over the whole Pacific . . . with a maximum off Baja California."[42]

In 2014, reports supporting this prediction were starting to come in, showing rising levels of radiation, the highest concentration at Surfer's Beach in Half Moon Bay, a surfer and beach-lover's haven near San Mateo.[43,44] Half MoonBay has never had a measureable snowfall, but after Fukushima, the Geiger counters could sound like castanets. The State of California, though, waved a red flag not at the danger but at the fear of it: ". . . recent tests by the San Mateo County Public Health Department and CDPH [the state health department] show that elevated levels of radiation at Half Moon Bay are due to naturally occurring materials and not radioactivity associated with the Fukushima incident."[45]

The smiley-face announcements that there was only fear to be afraid of masked moves suggesting otherwise. In 2014, the U.S. Department of Health and Human services ordered 14 million doses of potassium iodide, a guard against thyroid cancer, readied by February.[46]

NOTES

1. "Tokyo, Japan," *Associated Press*, March 20, 2011.
2. "Murky Past of Japan's Troubled Nuclear Industry Revealed," *The Independent*, March 25, 2011.
3. Tony Long, "March 9, 1945: Burning the Heart Out of the Enemy," *Wired*, Mar. 9, 2011.
4. See William Craig, *The Fall of Japan* (New York: Dell Publishing Co., 1967).
5. Richard Krooth and Hiroshi Fukurai, in *Common Destiny: Japan and the United States in the Global Age* (Jefferson, North Carolina: McFarland & Company, 1990), pp. 225–226.
6. Ernest J. Sternglass, "Low-Level Radiation" in *Low-Level Radiation* (London: Earth Island Limited, 1973), pp. 6–9.
7. Ernest J. Sternglass, "Leukemia and Ionizing Radiation," in *Low-Level Radiation* (London: Earth Island Limited, 1973), pp. 16–17.
8. Dr. John Goffman, interview.
9. Ibid; Sternglass, passim; Richard Krooth vs. University of Wisconsin, et al. (1970).
10. Ernest J. Sternglass, *Low-Level Radiation* (London: Earth Island Limited, 1973), p. 9.
11. Evan Douple, Why We May Not Learn Much New About Radiation Risks From Fukushima, interview by Richard Knox, March 24, 2011, http://www.npr.org/blogs/health/2011/03/24/134833008/why-we-may-not-learn-much-new-about-radiation-risks-from-fukushima.
12. Annika Breidthardt, "German Government Wants Nuclear Exil by 2022 At Least," *Reuters*, May 30, 2011.
13. BC Hydro, *New Act Powers B.C. Forward with Clearn Energy and Jobs* (2010).

14. L. Phillips, "Europe Divided Over Nuclear Power After Fukushima Disaster," *Guardian*, May 25, 2011.

15. Don Eichelberg, "Public Is Opposed, Industry Pushes On," *Nuclear Guardianship Forum*, No. 3., Spring 1994.

16. Toko Sekiguchi, "Japan, Vietnam to Move Forward on Nuclar Deal," *Wall Street Journal*, October 31, 2011.

17. "Japan's Energy Pact with Turkey Raises Nuclear Weapons Concerns," *Asahi Shimbun*, Jan. 7, 2014.

18. Bossew, Peter, et al, *35 Years Promotion of Nuclear Energy: International Atomic Energy Agency; A Critical Documentation of the Agency's Policy* (Vienna-Hannover: Gruppekologie,: AntiAtom International and kologie Institut., 1993).

19. Taylor, Theodore B., *Nuclear Weapons, Nuclear Power and Solar Energy*, 1996; Greg Ruggiero and Stuart Sahulka, "Critical Mass: Voices for a Nuclear-Free Future" (presented at the Open Media and The Campaign for Peace and Democracy, Westfield, NJ, n.d.).

20. The Department of Energy is allowed to let 10 millirem per hour of gamma radiation come through the (WIPP TRUPACT-II type of) transportation container—measured at about 3 feet. (The maximum dose rate at the surface of the containers inside the transportation container—drums, standard waste boxes, etc.—can be as high as 200 millirem per hour.) Remote-handled waste has to be shielded with lead because most of its radiation is in the form of beta particles and gamma rays and the containers can emit up to 1,000 rem per hour (measured at their surface).

21. John W. Goffman and Egan O'Connor, "The Law of Concentrated Benefit Over Diffuse Injury" (presented at the Committee for Nuclear Responsibility, San Francisco:, 1994).

22. BEIR (Biological Effects of Ionizing Radiation) Committee, 1989, *Health Effects of Exposure to Low Levels of Ionizing Radiation: BEIR* V (Washinton, D.C.: National Academy Press, 1990).

23. Sternglass, passim; Steve Wing et al., "Mortality Among Workers at Oak Ridge National Laboratory," *Journal of the American Medical Association* vol. 265, no. 11 (March 20, 1991): 1397–1402.

24. Martin Fackler, "Japan's Prime Minister Defends Handling of Crisis," *New York Times*, Apr. 30, 2011.

25. "20 Millisieverts for Children and Kosako Toshiso's Resignation," *Asian Pacific Journal: Japan Focus*, May 01, 2011, available at http://www.japanfocus.org/events/view/83.

26. Martin Fackler,, "Japan's Prime Minister Defends Handling of Crisis," *Tokyo*,, April 30, 2011.

27. See "Minamata Disease Archives," Frequently Asked Questions, Question 6, available at http://www.nimd.go.jp/archives/english/index.html.

28. Reiko Seki, Sociologist (Rikkyo University, Tokyo, Japan, n.d.).

29. "Without SPEEDI, Residents Fled in Direction of Radiation," *Asahi Shimbun*, Dec. 27, 2011.

30. "U.S. Forces Given SPEEDI Data Early," *Japan Times*, Jan. 18, 2012.

31. "Koriyama Shimin—Hoshanokenen no Gakkojido Hihan Uttae Ga Saibansho De Kyakka [Koreyama Citizens: Court Denial of Plea to Relocate Children Due to Radiation," *Voice of Russia*, Apr. 25, 2013.

32. Id.

33. Norimitsu Onishi and Martin Fackler, "Anger in Japan Over Withheld Radiation Forecast," *New York Times*, August 9, 2011.

34. Id.

35. Id.

36. "SPEEDI Yosoku 'Kohyo Dekinai' Monbusho Bunsho ni Kisai [SPEEDI Prediction Impossible to Disclose: Recorded in MEXT Documents]," *Kyodo*, Mar. 3, 2012.

37. Id.

38. "U.S. Sailors Sue Japanese Nuclear Plant Owner TEPCO," *Environmental News Service*, January 3, 2014.

39. Dave Boyer, "NSA Spying Sours Relations with Germany, Brazil," *Washington Times*, October 24, 2013.

40. Document 12, "Activation of Echelon Units," from *History of the Air Intelligence Agency, January 1–December 31, 1994*, Volume I (San Antonio, TX: AIA, 1995).
41. Matthew M. Burke, "Congress Wants Answers on Health Impacts of Japan Disaster Relief," *Stars & Stripes*, January 27, 2014.
42. "Fukushima Radiation Levels Will Concentrate in Pockets at Specific U.S. and Canada West Coast Locations," *Global Research News*, November 2, 2013.
43. Adam Salazar, "The Good and Bad News About West Coast Radiation," *Alex Jones' Infowars*, Jan. 14, 2014.
44. Id.
45. Jason Hoppin, "State Rebuffs Radiation Concerns at Beach," *Santa Cruz Sentinel*, Jan. 9, 2014.
46. Anthony Gucciardi, "Is the Government Stockpiling Iodide in Preparation for Fukushima Meltdown?" *Storyleak*, Jan. 1, 2014.

Chapter Four

Damning the Torpedos at Full Throttle

The genetic foundation for the evolution of plants and animals includes flexibile response to change, a great potential for survival and reproductions and even species variation.[1] However, these genetic talents have been sporadically lost in calamities such as collisions between Earth and asteroids or comets; the waxing or waning of the Sun's radiant energy and heat and consequent desertifcation or ice ages; and, now the mix of hubris and oligarchical attitudes seen in the drive to power and profit by promoters of the use and development of ionizing nuclear materials that may threaten life on the planet and even the planet itself.

Immediate risks are as plain to be seen as they are, by some, easy to deny or ignore. Denial, deception, and distraction adds up to amnesia. In the countries of its perpetrators, Hiroshima Day passes almost unnoticed. Radiation concerns seem esoteric: what percentage of a market- and media-besotted world can even define the word "minisievert"?

For most of this world, what happened in Japan stays in Japan, likewise with Three Mile Island or Chernobyl. "Socially responsible" and widely seen films treated the nuclear threat as melodrama or satire, as in *The China Syndrome* or *Dr. Strangelove*, showing the Earth exploded while Vera Lynn sings the World War II hymn to hope, "We'll Meet Again." Yes, in Japan highly radiated foods were banned for a time after the Fukushima meltdown, but the U.S. never suffered a shortage of sushi. In Japan itself, the government briefly advised parents in Tokyo and northward not to let children drink tap water, but, after all, bottled water and rice-based beers such as Asahi produced not far from Fukushima or Kirin, some of which is produced by Anheuser-Busch in the U.S., were increasingly and globally popular in the early twenty-first century. The sun never sets on Asahi (The Rising Sun) drinking, and its giant beer hall in Tokyo rivals the fame and surpasses the architecture of the Hofbrauhaus in Munich. Japanese beer, though 67 percent malt, is permitted, also, to contain 33 percent of additives such as rice, corn, sorghum, potato, starch and sugar, and no information is available about its extra-alcoholic safety.

FROM HERE TO ETERNITY VIA NUCLEAR POWER

As containment leaks haunt the nuclear industry, it has, with equal lack of success, tried to plug all seepage of dire facts into the awareness of the concerned. For example, when hundreds of thousands of civilian guinea pigs were blasted with direct or radiant power, President Truman called the mass murder a military necessity, because of the frantic defense that the Japanese would somehow pull together against a land invasion of their already-destroyed country—despite his administration's and the military's knowledge of Japan's offer to surrender, and its weakness exemplified in the clear skies and unopposed delivery of Fat Boy from the womb of *Enola Gay*.

Few even sought shelter in Hioshima and Nagasaki—what shelter was left? People thought the unescorted A-bomb planes were checking the weather, though their hit-men operators already knew that it would be turbulent, murderous, to a high degree.

Seventy to 100,000 people below were blasted into eternity in minutes. More were blown across the landscape like paper dolls in the wind or entombed in rubble, the voices of the damned still possessed of voices raised in pain or desiccation or fear or all of these. What reembled humanoid zombies, brain dead, shambled after their own lost flesh or flesh and blood. From these burned naked and dazed creatures' bodies hung down sooty flaps and folds. Most had singed and frizzled hair or bald, scarred skulls, burns black and swollen. Many lacked faces, were left only eye and nose craters. They vomited ceaselessly or stumblingly emitted wastes on themselves. Survivors near the initial sites died within three days of radiation poisoning.[2]

A recently disclosed document from intelligence units in U.S. Air force Headquarters says it had laid down criteria for municipal massacres, the first being cities that had military targets in them, which could include nearly every major city in Japan. Second, pilots were directed to areas which would show most clearly outcomes of the "experiment": level terrain and a variety of construction bases for the structures in the blast zone, so as to obtain reports on the efficiency of the new weapon. The Air Force Mengeles were inconvenienced, too, by the requirement that a target had to be one not previously visited by the 20th Air Force. "Call it a virgin target undamaged, unhurt by any other type of an explosive or munition."[3]

The aerial rape list of General Leslie Groves, commander of the bomb project, put Kyoto, Japan's ancient capital and shrine, near the top. However, a fluke saved that undefended treasure. "Secretary of War Henry Stimson, approaching 80 years of age, would not have it. He had visited Kyoto with his wife in the twenties and had enjoyed the city's cultural riches. It was a city of great religious significance to the Japanese, and Stimson felt Kyoto's destruction would damage America's post-war stature."[4]

A lifespan after the nuclear obliteration of Hiroshima and Nagasaki in these and other nuclear sites, soil was saturated with poison, and the radioactive wastes that postwar nuclearism produced spread toxins from every storage or processing site, a clean disposal or stockpile remaining impossible, despite elaborate and much publicized efforts.[5]

Dangers escalate. They breed. Fuel rods of enriched uranium used for three years in a reactor become approximately one million times more radioactive than when first loaded. The nuclear industry calls such rods "spent" fuel and is required by law to store them safely forever. Yet governments and nuclear industries for reasons of poltiics and profits continue to develop and use nuclear power, even though though they can only guess about the efficacy and permanence of waste safety measures.

Enriching uranium to create fuel rods and recycling reactor fuels also produces the waste byproduct depleted uranium (DU).[6] Spent fuels contain other highly radioactive substances, such as plutonium, neptunium, technetium, and U-236, whose concentrations vary. Plutonium, 200,000 times more radioactive than U-238, presents more serious dangers than other radioactive ingredients.

The U.S. government and nuclear businesses have also found out how to "recycle" nuclear wastes in the form of depleted uranium (DU) ammunition. Since 1991, U.S. armed forces have used such weapons in the bombing of cities and towns in Iraq, Bosnia, Vieques, Yugoslavia, Iraq, Afghanistan, Somalia, Yemen, Pakistan, and Libya. During the first Gulf War in 1991 alone, both the U.S. and British airplanes fired off more than a million rounds of DU-coated bullets, and in Kosovo, the Pentagon admitted to using 31,000.[7] After thousands of Gulf War veterans sickened, investigators found increased cancer rates in vets soldiering where they fired DU ammunition.[8] Of 700,000 U.S. veterans from the first Gulf War, greater than 240,000 are on pernament medical disability and 11,000 have died.[9] Another investigation also found that four of nine soldiers of the 442nd Military Police Campany of the New York Army National Guard returning from Iraq tested positive for DU contamination and became the first confirmed cases of inhaled DU exposure from the first Iraq war.[10] Nuclear waste was also used for the manufacture of bunker-busters, including the fifteen-ton Massive Ordinance Penetrator (MOP) designed to demolish weapons bunkers buried underground and shielded by 10,000 pounds of reinforced concrete.[11] The continuous use of DU weapons allowed the U.S. government to dump massive amount of nuclear wastes in many "enemy" countries.

Nuclear accidents and aerial toxics also pose uncontrollable risks. Ten days after the Chernobyl meltdown in Ukraine, the nearby radiation level was measured at fifty million curies, equal to the effect of exploding more than one hundred Hiroshima-size bombs.[12] Winds spread this radioactivity around the globe. Children of Belarus and Ukraine surviving the initial exposure developed extraordinary rates of thyroid cancer and leukemia; more others were listless and sickly and have memory difficulties and hemophilia. Ukraine reported that, as of 1985, the catastrophe had caused 125,000 fatalities.[13]

Chernobyl-caused cancer spread across Europe. The 2009 study published by the New York Academy of Sciences estimates that from 1986–2056, 89,851 will die from various Chernobyl-caused cancers.[14] The report categorized the Chernobyl dead and dying as (1) the cleanup crews (so-called "liquidators") sent in immediately after the April 1968 disaster; (2) the unnumbered dying from 1986 to 1989 during the Soviet-enforced silence; and (3) the 112,000 to 125,000 people from post-disaster efforts or 15 percent of those ordered into the area by the Soviets. The report also revealed that, in Russian/Ukraine contaminated territories, cancers accounted for 3.8 percent to 4.0 percent of all deaths from 1990 to 2004.[15] Statistics for the entire Eurasian fallout zone are unavailable, but radiation-induced diseases identified include iodine-131-caused lymphatic and thyroid system dysfunction; endocrine maladies; circulatory, cardiovascular, and immune ailments; and genital and reproductive deformities due to leukosis of fetal, utero and urogenital, and sperm and urogenital cancers.[16] Increased levels of cancers and other diseases were also documented for people in or near Harrisburg, Pennsylvania, the site of the Three Mile Island nuclear accident.[17] Observers note also flora and fauna abnormalities, an aftermath prompting the creation of a citizens' monitoring network. The citizens' network is also a comment on the fact of avoidance of any official comprehensive, health, and environment study of Three Mile Island.[18]

JAPANESE JEOPARDY: WAVES OF DISASTER

Japan's nuclear disaster is no respecter of national boundaries, nor is it likely to be unique. Explosions at its power plants had causative factors still present elsewhere: aged equipment and structures, vulnerability to geophysical shock, and the elusiveness of fuels and wastes. Of these, the public knew little or assumed they had been prepared for, but when the Trojan horse disgorged its agents of assasination, eleven of Japan's fifty-five nuclear power plants, nearly 20 percent of the country's electrical capacity, went down. Japan suffered an economic blow and by-blows from which it has not recovered.

Backups were swept aside. When the cooling power for the fuel rods failed, the reactors switched to backup diesel generators that the tsunami put out of business, as the tsunami flooded them. Unit four then tapped excess steam in the core to power a turbine and switched to battery power with only a short half-life. Japanese utility and government officials made all efforts to move another generator to the site to prevent a partial meltdown similar to that in Three Mile Island in 1979. On Saturday morning, March 12th, they claimed success in restoring power, but the International Atomic Energy Authority disagreed: temporary power only was restored from "mobile electricity supplies."[19] And some interior fuel rods remained exposed rather than submerged in water.

Later the same day a Japanese nuclear safety official admitted that continuing overheating made a meltdown possible.[20] Not to worry, said a government spokesman, drawing on hope rather than fact: a meltdown would not affect residents outside the evacuated area's six-mile radius. Meanwhile, Tokyo Electric said it was venting "slightly radioactive" steam and gas to relieve pressure building in the containment building at unit No. 1. The utility avoided mention of lethality but, on its Web site, guessed the spike was "assumed to be due to leakage of reactor coolant." The company said it did not think there was leakage of reactor coolant in the containment vessel "at this moment."[21] The extent, effect, or even location of the leak remained unclear.

The plant was obviously hot. So the obvious solution, at least for Tokyo Electric, was to dump water on it, tons of water from helicopters. This steamed into the air and ran away in rivulets. The operator of the Daiichi, or Unit 1, plant soon stated that his team had detected radiation levels 10 million times higher than usual in leaked water at Reactor Two, and heavy steam rising from the tsunami-battered facility. The radiation level of 1,000 millisieverts per hour in the turbine building was too dangerous to approach, so repair and cleanup draftees scrambled for the exits—"evacuated" in the reassuring language of Tokyo Electric.[22]

Guesswork prevailed; but clearly the main concern was the health of the operation rather than that of human life and the environment.

REACTOR GASTROINTESTINAL PROBLEMS

A containment building is one of the major safety, or fail-safe, mechanisms in a nuclear power plant. It surrounds the core of the reactor and is meant to contain accidental releases of gases from the core. The hot core has cooling water, but if its level drops, coolant turns to steam and is released through special valves into the containment building, nuclear experts said. That could cause an increase in pressure inside the sealed containment building itself, and ultimately force a release of gas and steam through filters meant to keep most, though not all, of the radiation inside the building.

There were also reports of elevated radiation levels inside the *control room* of that reactor unit. NISA said levels were 1,000 times the norm. The AP later quoted an official from NISA as saying that a measurement of radiation levels *outside* the plant was eight times as high as normal.[23] Even that level of radiation still posed little danger to residents, government nuclear experts said. They also said the release of steam and gas from containment buildings posed little danger.

But on the Tokyo Electric Web site, reports of radiation effects and control problems began to appear. A specialist was treating a worker for radiation exposure, coolant water levels ebbed despite the injection of coolant directly into the reactor and the aerial flooding. On March 13, the status of four reactors at Tokyo Electric's Daini Nuclear Power Plant 6 miles south of the Daiichi plant was reported "unclear," contradicting earlier reports that they had shut down. The power supply, on- and off-site, for cooling failed; dangers were unresolved.[24] Engineers raced to restore AC power to prevent a core meltdown, but soon it was clear they had completely failed.[25]

MACROWAVY GRAVY

Boiling water with a radioactive system is hardly the most inventive exercise our species has undertaken. The ancients discovered they could bring water to a boil by capturing lightening bolts or focusing the Sun's radiant heat on a water reflector. Using flash radiation to boil water has been detected among aboriginal Australians from 100,000 years back. Today's technologies, however, enable scientists to boil water with a radioactive torch, then attempt to control the fire with other experimental legerdemain or snuff the fire with seawater.

Fukushima, however, shook the scientists of the brave new world. At 1:30 p.m. EST on March 12 or 3:30 a.m. March 13 in Japan, U.S. nuclear experts joined a conference call on the crisis initiated by the media. Many speculated about what the accident would do to policy; physicist and nuclear accident researcher, Ken Bergeron addressed, rather pessimistically, what was actually happening:

> [This] type of accident is known as a station blackout, i.e., loss of offsite AC power and then a subsequent failure of emergency power on site . . . station blackout has been one of the great concerns for decades. . . . The probability of this occurring is hard to calculate primarily because of the possibility of what are called common-cause accidents, where the loss of offsite power and of onsite power are caused by the same thing. In this case, it was the earthquake and tsunami. So we're in uncharted territory, we're in a land where probability

> says we shouldn't be. And we're hoping that all of the barriers to release of radioactivity will not fail. . . . The fuel rods are long uranium rods clad in a zirconium alloy casing held in a cylindrical-shaped array. The water covers all of that. If the water descends below the level of the fuel, then the temperature starts rising and the cladding bursts, releasing fission products. Eventually the core begins slumping and melting.[26]

Former U.S. Nuclear Regulatory Commission (NRC) member Peter Bradford said "the other thing that happens is that the cladding, which is just the outside of the tube, at a high enough temperature interacts with the water. It's essentially a high-speed rusting, where the zirconium becomes zirconium oxide and the hydrogen is set free. And hydrogen at the right concentration in an atmosphere is either flammable or explosive."

"The danger is the very thermally hot reactor cores at the plant must be continuously cooled for 24 to 48 hours," said Kevin Kamps, a specialist in nuclear risks. "Without any electricity, the pumps won't be able to pump water through the hot reactor cores to cool them."[27]

THE MOST VENOMOUS SUBSTANCE ON EARTH

Aside from human evil, a topic beyond but perhaps suggested by this discussion, the world's most posionous substances are thought to flow from the fangs of reptiles like the black mamba, which injects over ten milligrams of poison into its victims. The lesser-known beaded sea snake can kill six people with a few drops of its venom. Such toxicity, however, pales in comparison with an almost weightless and naturally undetectable few microns of plutonium. This substance is fuel and product of nuclear reactors. Unlike snake venom, used for purposes of defense and prey, the deadlier element merely comes into play to bring liquid to a simmer.

Put another way, the Fukushima Daiichi reactors are all General Electric–designed boiling-water reactors. Such reactors, bunkered, send steam through piping into turbines. These convert the steam's energy into a generator whose turning produces electricity. This simple process produces power to run the industrial machinery of what became known as an energy alternative to fossil fuels like gas, oil, and coal. Instead of wood, water, coal, or oil, the heat required is from the energy released by the splitting of uranium or plutonium atoms. When split, the atoms lose mass, which is converted to energy. Some of the energy may escape the system under normal conditions; nearly all in catastrophic episodes—thus mortal danger always lurks near nuclear generation.[28]

Linus Pauling wrote in 1975, "I advocate that no more nuclear fission power plants be constructed, and that instead we concentrate our efforts on . . . developing supplies from inexhaustible resources. The immediate step to be taken is that of stopping the construction of nuclear fission power plants."[29] His and other voices cried in a wilderness, a gold rush of building, selling, and acquiring systems for sending water through nuclear reactors, for acquiring infinite energy, too cheap to meter, and producing interesting by-products that could be used for weaponry and war.

Production obsession deflected cares about containment, proper maintenance, or environmental protection. To take one example, hydrogen volatility and pressure, meant to be controlled by scrubbing and valve adjustment, could exceed safe limits and perhaps notice. Used in the turbine hall to cool the electric generator and within the reactor, it sometimes escaped and exploded. Then, in more serious cases, as at the Daiichi reactor, hydrogen could react with the metal of the fuel rods to produce gaseous steam. At Daiichi, operators may even have vented hydrogen into the turbine hall, actually causing the explosion.

Something was leaking, certainly, or mishandeld. The day of the explosion, March 12th, a Japanese nuclear safety panel reported that the radiation levels were 1,000 times above normal in the reactor *control room* at Daiichi and radiation levels *near the main gate* eight times normal levels, according to Japan's NHK radio-TV reports.[30]

To sum up the Daiichi Plant crisis, trouble began soon after the earthquake Friday afternoon on the 11th. Electric power went out; emergency diesel generators kicked in to run cooling but shut down about an hour after the earthquake. The tsunami possibly had flooded and disabled the generators. Foreign experts, media, and the public could only guess about causes, but the effects were very clear. A disaster commenced.[31]

MELTDOWN MECHANICS

On Saturday March 12th, as operators desperately tried to prevent Unit 1 from overheating and melting down, an explosion shattered its walls. With no power, and valves and pumps damaged by the tsunami, Unit 1 authorities resorted to drawing seawater mixed with boron to try to cool the overheating uranium fuel rods. Boron is constituted to disrupt nuclear chain reactions, but the move likely rendered the forty-year-old reactor unusable, said a foreign ministry official. Officials also said that the seawater would remain inside the unit, possibly for several months.[32]

Robert Alvarez, senior scholar at the Institute for Policy Studies and former senior policy adviser to the U.S. Secretary of Energy, told reporters that using seawater was a long shot. The success of using seawater and boron coolant would depend on the volume and rate of its distribution. He also said the dousing would need to continue nonstop for days. Another key would be restoring electrical power to bring back to life the normal cooling systems.[33]

Reportedly, the cooling operation at Unit 1 went smoothly after the seawater was pumped in. And operators released slightly radioactive air from Unit 3 on Sunday, the 13th, while injecting water into the unit, hoping to reduce pressure and temperature to prevent a possible meltdown.

The developing catastrophe was a mix of high drama and charade. Japan's nuclear agency warned that the reactor, despite operator efforts, could be nearing a meltdown and that two radioactive substances, cesium and radioactive iodine, had been detected nearby. A government spokesman said radiation levels just outside the plant rose briefly past legal limits but had since declined significantly. He said fuel rods were exposed briefly, indicating that coolant had not covered the rods for some time. Their exposure contributed further to raising the temperature in the reactor vessel.

By Sunday, in Unit 1 the massive explosion that crumbled the walls of the reactor building had wounded workers struggling to gain control of the quake-stricken unit. The blast came when hot vapor from the reactor's steel container released hydrogen that mixed with outside oxygen and blew off the roof and walls of the building around the containment vessel, but allegedly left whole the container itself. White smoke billowed into the air. Japanese officials warned those nearby to cover their mouths and stay indoors.[34] At 7:00 p.m., Daiichi plant general manager Masao Yoshida ordered seawater injection into reactor 1, and ignored a Tokyo Electric order at 7:25 to stop such infusions.[35]

The public was clueless and fearful. Before the explosion, the Japanese government herded people 1.9 miles (3 km) from the Daiichi plant. Those within a radius of 6.2 miles (10 km) were told that they could remain in their homes and carry on regular activities.[36] After the explosion, the government widened the evacuation zone to a 12.5-mile (20 km) radius from the plant and prepared to distribute iodine tablets to people in the vicinity to counter radiation effects.

On March 14, at 11:01 a.m., another building housing the Unit 3 reactor exploded, injuring six workers. Tokyo Electric instantly declared there was no release of radioactive material beyond that already being vented, but the blast damaged the water supply to Unit 2.[37] On March 15, at 6 a.m., another explosion occurred at the fourth floor rooftop area of the Unit 4 reactor.[38] Contradicting the the utility company's press release that said the structure was still basically sound, aerial photos showed most of the outer building in ruins.[39] The World Meteorological Organization (WMO) also reported a second explosion of the Unit 3 reactor at 11:00 a.m. on March 15.[40] The WMO also stated that many surrounding areas including Tokyo, Nagoya, Sendai, and other metropolitan regions were detecting high levels of radiation due to the easterly winds and precipitation in and around Fukushima.[41] A fire also broke out at the Unit 4 and radiation levels at the plant rose significantly but subsequently fell, according to Tokyo Electric.[42]

REPEAT PERFORMANCE: TRIPLE TROUBLE

The explosions at the Daiichi plant rattled confidence in nuclear power in a country long sensitized to the dangers of radioactivity. The Japanese government declared a state of emergency for five reactors at two nuclear power complexes, as military and utility officials scrambled to tame rising intra-unit pressure and radioactivity levels and to steady reactor core cooling systems.

On March 12, Tokyo Electric admitted rising temperatures and pressure showed it was having trouble controlling three of the four reactors it owns at nearby Fukushima Daini plant, some six miles (10 km) south of the Daiichi plant.[43]

Then the situation turned for the worst. As the nation grappled simultaneously with its worst nuclear accident and the aftermath of its largest recorded earthquake, Japanese officials took the bizarre step on March 12th to try to avoid meltdown by flooding with seawater and completely destroying Daiichi's crippled nuclear reactor.

As we have seen, officials had been venting radioactive steam into the public domain near Fukushima Daiichi's Unit 1 to relieve pressure inside the reactor vessel housing overheating uranium fuel. Concerns escalated dramatically on March 12th when that unit's containment building exploded. NISA spokesman Shinji Kinjo stated that officials were aware that the steam contained hydrogen and were risking an explosion by venting it, but chose to do so because they needed to keep circulating cool water on the fuel rods to prevent a meltdown. But officials insisted there was no significant radioactive leak after the explosion.

It was a bald-faced lie, dangers to the surrounding population escalating by the hour. If a full-scale meltdown got going, melted fuels would eat through the bottom of the reactor vessel, then through the floor of the containment building. At that point, the fissionable uranium and dangerous byproducts would escape into the environment, with potentially deadly effect on all life. Eventually, the walls of the reactor vessel—measuring six inches (fifteen centimeters) of stainless steel—would melt into a lava-like pile, slump into any remaining water on the floor, and potentially cause another explosion that would spash radioactive poison farther.

As the destruction and spewing went on, Peter Bradford, former commissioner of the U.S. Nuclear Regulatory Commission, told reporters that dangers would multiply if the reactor core became exposed to the outside.[44] At that point, Fukushima officials would likely begin pouring cement and sand over the entire facility, as was done at the 1986 Chernobyl nuclear accident in Ukraine. Physicist Ken Bergeron also said such a meltdown would put the surrounding area off-limits for as much as 25,000 years, and "a lot of first responders would die."[45]

On the 12th, the cooling system at a second reactor at Fukushima Daiichi failed, forcing officials to release more radioactive vapor into the atmosphere. And some 210,000 people were told to evacuate the area instantly.

On March 13, at reactor No. 3, cooling system failure exposed fuel rods, a first step to meltdown. So operators again tried injecting water and boric acid into the reactor and eased pressure by venting radioactive vapor, said Yukio Edano, the chief cabinet secretary. He said extraordinary measures to disable the reactor were required to cool the exposed rods.[46] He said nothing about the irradiation of surrounding populations, many on the road and more in the exclusion zones highly exposed, but as he spoke radiation levels outside the plant, which had retreated overnight, shot up to 1,204 microsievert per hour, or over twice Japan's legal limit.

DANGEROUS SAGA CONTINUES

To understand the radiation danger in Fukushima, one should take into account reactor design, a so-called fail-safe make-up introduced by General Electric decades back. Unlike newer structures, Daiichi's GE-reactors included dozens of holes pierced in the bottom of each reactor vessel. Each hole allows one control rod to slide into the reactor from below. The control rod itself was made of a neutron-absorbing material that could quickly stop nuclear fission inside the reactor. During normal operations, a graphite stopper covered each hole, sealing in highly radioactive primary cooling water. But at temperatures above 350 degrees Fahrenheit, the graphite stoppers begin to melt—opening the gates to a flood of failure of technology.

When the earthquake and tsunami cracked the fuel rods, the consequent rubble collected at the bottom of the reactor, and heat and radiation jumped and ate away the seals. As the graphite seals failed, extremely radioactive reactor water leaked into a network of ruptured pipes, from there into the reactor turbine containment buildings—the very buildings that had already been heavily damaged by explosions.

Leaked water sampled from the Unit 1 reactor was 100,000 times more radioactive than normal background levels, somewhat better than Tokyo Electric's first faulty calculation which had put it and water from Unit 1 as being 10 million times higher than normal. At any rate everyone in both units scrambled for exits, some of which were blocked.[47]

Even after Tokyo Electric "correctly" measured lethal levels, airborne plutonium radioactivity in the Unit 2 turbine building still remained so high—1,000 millisieverts per hour—that a worker there would reach his yearly occupational exposure limit in fifteen minutes. About half of the workers exposed to a dose of 4,000 to 5,000 millisieverts absorbed as they worked and then ran would die of radiation poisoning. Hoping to—and told they could—avoid this fate, people in the danger zones walked off their posts and were replaced every thirty to sixty minutes.[48]

Outside the plants, however, dangers multiplied. Radioactive water flooded four of Fukushima's six turbine rooms and engineers had no quick way to clean it up. To stabilize the facility, workers operating in dark, steamy conditions without guidance continued trying to repair the elaborate cooling system designed to keep reactor cores and spent fuel ponds from overheating. Wearing respirators, face masks, and bulky suits, they raced to repair motor pumps as large as automobiles, in an unlit cavern of cables. To the radioactive fog of the facilities was added a miasma of fear settling on anyone in or near the half-dead nuclear utilities.

Reports piled up of damage to equipment and structure from the quake and tsunami. Saltwater poured over the reactors in the vain attempt to cool or kill them had corroded connectors, piping, seals, and equipment. Nuclides were pouring and aereated from the plants. Japan's nuclear agency reported on March 27 that a new measurement of seawater about 1,000 feet from the facility showed an iodine-131 level 1,850.5 times the legal limit, higher than a reading taken the previous day.[49]

AN IMPOSSIBLE MISSION

Unable to control the realities, Tokyo Electric and the govrnment tried to manipulate perceptions. They discounted or denied every update on the crises at the Daiichi plant, or tried to ameliorate the reports by saying they contained error or were transient findings.

On March 27, there was deep concern about initial readings of radioactive iodine-134, which has a half-life of only fifty-three minutes and would not be present in large quantities unless nuclear fission had restarted, presenting the alarming possibility of an out-of-control reactor meltdown. Several hours after releasing the initial results, the plant operator said that those readings had been in error, and that retesting had shown negligible amounts of the isotope. Yet the Unit 2 reading on a dosimeter had jumped past the highest reading, 1,000 millisieverts per hour. Michiaki Furukawa, nuclear chemist and board member of the Citizens' Nuclear Information Center in Tokyo, noted that 1,000 millisieverts of radiation induced nausea and vomiting, and higher amounts were possibly lethal.[50]

Still there was no evacuation of the roughly 1,000 workers, many assigned by the so-called Yakuza or Japan's organized criminal syndicates, still stationed at Daiichi after such high radiation levels were discovered. Naoki Sunoda, a Tokyo Electric spokesman, said that since the crisis began, nineteen workers had been exposed to radiation levels of 100 millisieverts.[51] The closely controlled workers, many of them in debt to the Yakuza and somewhat reassured by the shift system, kept at it, trying to find a way to approach the turbine building of reactor No. 2 to bleed off contaminated water. Sunoda implied that normalcy was just around the corner: "our objective is to restore power to all reactors so cooling functions can be restored."[52] Not around the corner but coming closer down the road were death and contamination, first for the brave workers assigned by Tokyo Electric and workers drafted by the Yakuza and later for millions of Japanese and miles of environment, if not greater quantities of people and place.

GANGLAND GYPSIES BECOME NUCLEAR FODDER

After the March 11 disaster, Tokyo Electric implored help from major construction and maintenance companies. These firms in turn hired workers from a sprawling network of six hundred contractors of varying expertise and reputation.[53] The need, and the complexity of Fukushima contracts and the shortage of workers played into the hands of the Yakuza, long expert in labor racketeering. Nearly fifty Yakuza gangs operated in Fukushima Prefecture, dominated by three major syndicates, i.e., Yamaguchi-Gumi, Sumiyoshi-Gumi, and Inagawa-Kai. Reports of Yakuza involvement, shoddy labor, and delegation of responsibilities to inexperienced nuclear cleanup subcontractors were commonplace.[54]

Japan's nuclear industry relied on cheap labor since the first plants including the Unit 1 of Fukushima opened forty years ago. The industry and its enforcers rounded up itinerant workers or debt-muddled jobless called "nuclear gypsies" from urban slums and centers for the homeless, such as the Sanya neighborhood of Tokyo and Kamagasaki in Osaka. These took on low-paying but dangerous jobs such as construction and maintenance of nuclear plants. The Fukushima disaster boosted such wage slavery, not only creating demand but serving as cause to loosen previous regulations for employers. For example, when parliament approved funding decontamination in August 2011, the law failed to apply existing rules regulating the construction industry. Contractors were not required to disclose information on management or undergo any screening, suggesting that anyone could become a nuclear contractor overnight. Small companies historically linked with the Yakuza syndicates rushed to bid for contracts and turned to human traffickers to round up the manpower, much of it as untrained and ignorant of what they faced as volunteers of 1914 headed for trench warfare.

NOTES

1. Sean B. Carroll, *Endless Forms Most Beautiful: The New Science of Evo Devo* (New York: W.W. Norton, 1995).

2. National Academy of Science, *Biological Effects of Ionizing Radiation*, February 19, 1998.

3. "Atomic Cover-Up: The Hidden Story Behind the U.S. Bombing of Hiroshima and Nagasaki," *Democracy Now*, Aug. 9, 2011, available at http://www.democracynow.org/2011/8/9/atomic_cover_up_the_hidden_story.

4. Id.

5. Nicholas Lenssen, *Nuclear Waste: The Problem That Won't Go Away* (Washington, DC: Worldwatch Institute, 1991).; Goffman *passim*.

6. "'Depleted' Uranium Munitions: Nuclear Waste as a Weapon," *Military Toxics Protect: Information Sheet*, June 2003, available at http://www.dmzhawaii.org/wp-content/uploads/2009/02/depleted-uranium-fact-sheet.pdf.

7. "Former U.S. Army Depleted Uranium Expert Alleges Campaign of Harassment Against him for Speaking Out," *Democracy Now*, May 30, 2000, available at http://www.democracynow.org/2000/5/30/former_u_s_army_depleted_uranium .

8. Id.

9. Thomas Weiss, "Dangers and Health Effects of Depleted Uranium," *Disabled World*, Sep. 4, 2013.

10. Juan Gonzalez, "Army to Test N.Y. Guard Unit," *NY Daily News*, Apr. 5, 2004.

11. "Ultimate Bunker Buster: U.S. Military Speed Plans for 13-Ton Bomb," *Daily Mail*, Oct. 13, 2009.

12. Dr. John Goffman, Interview of Dr. John Goffman, Diablo Nuclear Plant, CA, interview by Richard Krooth, 1976.

13. Michael Mariotte, "On Ninth Anniversary, Ukraine Officials Say 125,000 Have Died From Chernobyl," *Nuclear Monitor, Washington, DC* vol. 10, no. 5, Nuclear Information and Resource Service (May 1995).

14. Alexey Yabloskv, Vassily Nesterenko, and Alexey Nesterenko, *Chernobyl: Consequences of the Catastrophe for People and the Environment* (New York: The New York Academy of Sciences, 2009).

15. Ibid.

16. Ibid.

17. M.C. Hatch et al., "Cancer Near the Three Mile Island Nuclear Plant: Radiation Emissions," *American Journal of Epidemiology* vol. 132, no. 3 (1990).

18. Dr. John Goffman, Interview with Dr. John Goffman, Offices of Atty. Melvin Belli, San Francisco, CA, interview by Richard Krooth, 1976;

19. Steven Mufson & William Branigin, "Explosion Rocks Japanese Nuclear Power Plant: 5 Reactors in Peril," *Washington Post*, March 11, 2011.

20. Ibid.

21. Ibid.

22. Shingo Ito, "Radiation Levels Had Soar at Japan Nuclear Plant," *Agence France-Presse*, March 11, 2011.

23. Mufson & Branigin, supra note 19.

24. Steve Mirsky, "Nuclear Experts Explain Worst-Case Scenario at Fukushima Power Plant," *Scientific American*, March 13, 2011.

25. Steve Mirsky, "The Japan Earthquake and Tsunami," *Scientific American*, March 12, 2011.

26. Mirsky, Supra note 24.

27. William Pentland, "Risk of Nuclear Catastrophe Escalates in Japan—Worse than Chernobyl," *Forbes*, March 11, 2011.

28. John J. Berger, *Nuclear Power: The Unviable Option* (Palo Alto, California: Ramparts Press, 1976).

29. Ibid.

30. Hiroko Tabuchi and Matthew L. Wald, "Japanese Scramble to Avert Meltdowns as Nuclear Crisis Deepens After Quake," *New York Times*, March 12, 2011.

31. Ibid.

32. Scott Baker, "'Hail Mary Pass' Japan's Frantic Efforts to Avoid Nuclear Meltdown," *Blaze*, Mar. 12, 2011.

33. Ibid.
34. "Reactor Core Status of Fukushima Daiichi Nuclear Power Station Unit 1," *TEPCO*, May 15, 2011.
35. Ibid.
36. "Timeline: Japan Power Plant Crisis," *BBC News*, March 13, 2011.
37. "White Smoke Around the Fukushima Daiichi Nuclear Power Station Unit 3," *TEPCO*, March 14, 2011.
38. "Damage to the Unit 4 Nuclear Reactor Building at Fukushima Daiichi Nuclear Power Station," *TEPCO*, March 15, 2011.
39. "High-Resolution Photos of Fukushima Daiichi," *Air Photo Service Co. Ltd., Japan*, March 20 and 24, 2011.
40. "Summary Report of RSMC Beijing on Fukushima Nuclear Accident Emergency Response," World Meteorological Organization, October 31, 2011.
41. Ibid.
42. "Container Damaged, Radiation Leak Feared at Fukushima No. 2 Reactor," *Kyodo News*, March 15, 2011.
43. Steven Mufson, "Japanese Nuclear Reactors in Peril, Radiation Surges After Earthquake, Tsunami, *Washington Post*, Mar. 12, 2011.
44. Eric Talmadge and Mari Yamaguchi, "Death Toll Tops 10,000, Multiple Nuclear Meltdowns Loom," Associated Press, Mar. 13, 2011.
45. Ibid.
46. "Partial Meltdown in Japanese Nuclear Plant?" *Mystical Poilitics*, March 13, 2011.
47. Chico Harlan and Brian Vastag, "Radiation Levels at Japan Nuclear Plant Reach New Highs," *Washington Post*, March 27, 2011.
48. Ibid.
49. Ibid.
50. "Soaring Radiation Spikes in Water At and Around Fukushima Daiichi Hindering Progress to Cool Reactors," *Bellona*, Mar. 28, 2011.
51. David Jolly, Hiroko Tabuchi, and Keith Bradsher, "Tainted Water at 2 Reactors Increases Alarm for Japanese," *New York Times*, March 27, 2011
52. Ibid.
53. Justin McCurry, "Fukushima Cleanup Recruits 'Nuclear Gypsies' From Across Japan," *Guardian*, Jul. 13, 2011.
54. Ibid.

Chapter Five

Facts about Nuclear Reactors Your Mother Never Told You

The type and extent of failure in nuclear reactors has seemed both incomprehensible to the public and unstoppable by nuclear experts. Aside from abstractions such as "nothing to fear but fear itself," the Japanese government was left holding the bag, or Pandora's box, when the lid blew. It was besides, prodded by Tokyo Electric agents in and out of the nationalistic regime to avoid a meltdown in public regard for nukes. The word had spread a few weeks after the flooding, explosions, and venting that plutonium-laced lakes were rising inside the three reactor buildings of Daiichi and was progressing through tunnel labyrinths and nearing the tops of their vertical openings, just two hundred feet from the sea.

Reactor 3's sea-bound surface water clocked in at more than one thousand millisieverts an hour, twenty times the maximum annual exposure for American nuclear plant workers. It could not only pollute but eat into piping and increase leaks of radiation detected in the seawater around the plant and in tunnels and ponds at the No. 1 and No. 2 reactors.

In mid-March, the Institute for Energy and Environmental Research (IEER), an independent nonprofit group in Takoma Park, Maryland, said the plant had already released far more radioactivity than the 1979 Three Mile Island accident. The institute warned that by March 22, the Fukushima reactors had released about 160,000 times the best estimate of the amount of iodine-131 from Three Mile Island, and about 10 percent of the amount released in Chernobyl in 1986. Fukushima Daiichi had released 2.4 million curies of iodine-131, as well as half a million curies of cesium-134 and cesium-137, which have longer half-lives. Those were also about 10 percent the amount released in Chernobyl. Arjun Makhijani, institute president, said "while the releases are still considerably below Chernobyl, they have already reached a level that could affect the region around the site for a prolonged period."[1] The word "period" for some meant for months, for others it meant for years. Signs indicate now that it may work out to mean for centuries.

On March 27th, Tokyo Electric acknowledged that it had plutonium in the plant's soil but lied about its dangers: the plutonium amounts found had a half-life of 24,000 years, but the company said it was no danger to humans. Scientists knew it would be a threat to generations yet unborn. Government spokesman Yukio Edano, said "this quake, tsunami and the nuclear accident are the biggest crises for Japan" in decades, and Premier Kan said he had sharply rebuked the plant's operator. "It's undeniable their assumptions about tsunamis were greatly mistaken," Mr. Kan justified his own initial inaction.[2] Both men intimated that the government itself was heroically determined to overcome this new Godzilla.

Further analysis produced more official duck and cover tactics. Plutonium found in the soil near the plant was evidence that the fuel in one or more reactors had undergone partial meltdown. The authorities said they had been assuming such to be the case in Reactors 1 and 2.[3] They claimed not to know the plutonium source, even suggesting that it was atom bomb test residue, as the gun still smoked: the highest soil levels of the poison in the soil were found about five hundred yards from the most heavily damaged reactors.[4]

WHO IS READING YOUR READINGS?

Despite the reports of leaked plutonium, the United States Agency for Toxic Substances and Disease Registry issued its finding that all readings were within the safe range of plutonium levels in sediment and soil. Tokyo Electric had already said that the highest reading was over three times the average of the past twenty years.[5] All three kinds of nuclear fuel at the complex could leak plutonium. Most reactor fuel was uranium with plutonium as a byproduct, but Reactor No. 3 was using mixed oxide fuel, or MOX, composed of plutonium and uranium. NISA deputy director general Hidehiko Nishiyama said that while the detected plutonium levels were low, they pointed to the gravity of the crisis. Plutonium is a heavy element and not easily dispersed, he said, so higher levels outside the reactor suggested a wider leak: "This detection from the soil speaks to the importance and seriousness of the situation."[6]

RUBE GOLDBERG DESIGN, ATOMIC TECHNOCRACY'S PRESSURE POINTS

Technical dysfunctionality popped up everywhere at Fukuyama, endangering not only the Japanese, but all reached by its cloud of radiation. Enormous discharges of nuclear heat warmed the air, loading the atmosphere with nuclides able to modify genetic structures of animal and plants. The nuclides shot holes in living matter, penetrating cellular integrity, morphing both ova and sperm cells, leading to reproductive disorders and congenitally malformed fetuses.

The following shows a rundown of the complex mechanics and multiple factors giving rise to lethal bases of danger in or near, or far beyond, a reactor.

(1) Using Fuel including MOX

With the fuel (Pu-239) being changed every three years, about half of the fuel is "burned" in a nuclear reactor, furnishing about one third of plutonium's total potential energy.[7] Worldwide, seventy tons of plutonium in used fuel is removed each year when refuelling reactors. The plutonium (and uranium) in used fuel is then recovered through reprocessing, as at Japan's Rokkasho facility. The plutonium can then be used with uranium in the manufacture of mixed oxide (MOX) nuclear fuel for electricity generation. But the generatiors using this MOX fuel lack the structural integrity to handle its super charge, leading to periodic malfunctions as at the Daiichi Unit 3 and elsewhere, where radioactive fuels are still in use.

(2) Fuel Damage

If pumps circulating a reactor's cooling water are not operating, the water in the reactor vessel will boil and evaporate, and in it the fuel can become exposed. The exposure also can follow a pipe leak or rupture that permits water to drain from the reactor. At this point, the zirconium cladding on the fuel rods will heat up, blister, then rupture. Radioactive material that normally collected in the gap between the fuel pellets and the cladding is then released into the reactor vessel.

(3) Radioactivity Release Pathways

The radioactive material released with fuel and coolant into the reactor vessel can stream into the primary containment. To protect it and attached piping from rupturing due to high pressure, relief valves automatically open to discharge steam—and the radioactive material along with it—into the primary containment building. Workers may also open the relief valves manually to vent pressure in the reactor that blocks the in-flow of makeup water, such as the seawater reportedly injected into Japanese reactors.

In addition, a steam-driven emergency system called the reactor-core-isolation-cooling (RCIC) system uses steam from the reactor vessel to spin a small turbine connected to a pump that transfers makeup water to the reactor vessel. After the steam is used for this purpose, the steam—along with the radioactive material—is pushed into the primary container.

(4) Partial Meltdown

Fuel uncovered by water and exposed for hours will start to melt. This makes cooling more difficult as the melted fuel clogs the spaces between the fuel rods. Melted fuel will collect on the bottom of the steel reactor vessel, sometimes called the lower head. It will begin to burn through the reactor vessel.

(5) Complete Meltdown

If the water level is low enough, most of the fuel in the core can melt and fall to the bottom of the reactor vessel. In hours the fuel melts through the steel bottom and onto the concrete floor of the primary container. This structure is designed to keep melted fuel and its radioactive rays within the reactor, but a meltdown can make it permeable, a special concern with the design of the Mark I containment vessels in use in Japan. The emissions will generate increasing pressure over a period of days and weeks, which can, if not relieved, collapse the whole containment building.

The primary containment has a typical leak rate of roughly 1 percent of its volume per day, so some of the radioactive material will leak into the secondary containment in the reactor building. The reactor building is normally kept at a reduced pressure relative to the outside so any radioactive emissions will stay inside the building. However, this requires the building itself to be intact. A meltdown does not necessarily mean a large release of radioactivity. This depends on the strength and integrity of the primary—steel—and secondary—concrete—containments.

(6) Radioactive Isotopes

Radioactive materials decay or release tissue- damaging and carcinogenic particles. Some elements have different forms, called isotopes, that differ in the number of neutrons in the nucleus. The radioactive isotopes of greatest concern in a nuclear power accident are iodine-131 and cesium-137. Iodine-131 has a half-life of eight days, meaning half of it will have decayed after eight days, and half of that in another eight days, etc. Therefore, it is of greatest concern in the days and weeks following an accident. It is also volatile so will spread easily on soil, by wind, or conveyance on clothes.

In the human body, iodine is taken up by the thyroid, and becomes concentrated there, where it can lead to thyroid cancer in later life, sooner in children exposed to iodine-131 than adults. To mitigate absorption of iodine-131, people may injest nonradioactive potassium iodine pills hindering the thyroid's absorbtion of iodine-131.

Cesium-137 has a half-life of about thirty years, so will take more than a century to decay by a significant amount. Living organisms treat cesium-137 as if it was potassium, and it becomes part of the fluid electrolytes and is eventually excreted. Cesium-137 is passed up the food chain. It can cause many different types of cancer.

Plutonium, a by-product of nuclear fission technology, with a half-life of 24,000 years, is the deadliest substance ever made by humans. Since it is so toxic and can be used to make nuclear weapons, plutonium-fueled reactors are also an extraordinarily dangerous way to generate electricity.[8] Plutonium is the most dangerous of all, nuclear power reactors producing a mixture of plutonium radio-nuclides. There is no doubt that plutonium, deposited in the human lung, is a powerful producer of lung cancer. Approximately five millionths of a single gram of reactor plutonium deposited in the lung will do it.

All nuclear power reactors, uranium- and plutonium-fueled, produce waste materials containing plutonium and other radioactive substances that can be used to make nuclear bombs. Thus every country that has nuclear power is a potential nuclear weapons state.[9] They exist in all parts of the globe and are increasing, some secretly.

TINKERING WITH DOOMSDAY

Fabricating a nuclear bomb is no secret, its most difficult challenge merely obtaining plutonium. Nuclear power reactors—437 in 31 countries, now all potential nuclear weapons states—have produced four-fifths of today's plutonium.[10]

Japan is acquiring one of the world's largest so-called civilian stockpiles of plutonium by shipping its spent fuel to France and Britain for reprocessing into plutonium. Japan thus could momentarily become a major nuclear weapons power,[11] and independent of others because of its completion of an enormous Nuclear Fuel Cycle Facility in Rokkasho Village in Aomori Prefecture on Northern Honshu, Japan. By 1997, this "one-stop shopping" nuclear site already included two uranium-enrichment-plants and high- and low-level nuclear waste storage facilities. A plutonium reprocessing plant was completed at the end of the twentieth century.[12]

Other countries have also attempted to develop a new type of commercial Fast Breeder Reactors (FBR), including the U.S., Britain, Germany, and Russia. However, all of them have abandoned the development. The breeder reactor is a new type of nuclear reactor designed to create more fissile materials than it consumes. Only France is still trying to repair its accident-prone Superphénix FBR. Yet, in a hopeful change of attitude, in October 1996, French economic analysts began publicly criticizing the project's enormous financial waste. They called the FBR "a grand failure of 60 billion Francs" ($12 billion) and recommended a serious look at a complete shutdown of the Superphénix before billions more were wasted.[13]

Japan's prototype breeder reactor "Monju" was built in Tsuruga, Fukui Prefecture, in Central Honshu and went online in 1994. However, the Kobe earthquake of January 1995 cracked its pipelines, 150 miles from the epicenter, and the reactor has been quiet since April 1995.[14] It was reported in February 2014 that the government will scrap plans to generate electricity at Monju.[15]

The reprocessing of used reactor fuel also creates highly radioactive wastes. Even the nuclear industry estimates that the quantity of wastes requiring long-term isolation from the environment is increased nearly ten-fold as a consequence of reprocessing.[16] Finally, all nuclear power reactors and places where quantities of radioactive material are stored could be as dangerous as nuclear weapon explosions. If they are not submerged in water coolant, or become the object of terrorist activity, military bombardment, or sabotage, radioactive material could spread on a vast and devastating scale.[17]

EARTH AS A NUCLEAR DUMP

Countries with commercial or military nuclear reactors lack a solution to the nuclear waste problem. Some keep wastes in "temporary" storage facilities or buried in shallow pits. A few have found a way to "recycle" them into depleted uranium ammunitions and bunker busters. Much more waste has simply been dumped directly into the ground, lakes and oceans of the world, including for example, the Irish Sea near Sellafield, England; the Pacific Ocean near the Farallones Islands off San Francisco, California; and Lake Karachay, near Chelyabinsk, Russia. Russia also dumped several decommissioned naval nuclear reactors in shallow waters in the Kara Sea in the Arctic Ocean north of Siberia. These included four submarine reactor compartments dumped in the Abrosimov Gulf in 1965 and 1966; three reactors from the nuclear-powered icebreaker *Lenin* into the Sivolky Gulf in 1967; another submarine reactor in the Kara Sea in 1972; Submarine K-27 in Stepovov Gulf in 1982; and another nuclear reactor in Techeniya Gulf in 1988.[18]

The U.S. dumped low-level nuclear wastes in the ocean from the late 1940s to 1970. But in 1970, the Council on Environmental Quality recommended that the seas be used only as a last resort.[19] In 1982, Congress passed the Nuclear Waste Policy Act (NWPA) directing the Department of Energy (DOE) to build and operate a repository for used nuclear fuel and other high-level radioactive waste. Congress in 1987 amended the NWPA and directed the DOE to study Nevada's Yucca Mountain as the site for a potential repository for geologic disposal of used nuclear fuel. In 2008, after two decades of site studies, the federal government filed a construction license application for a repository there, in a site held sacred site by Indian nations, including the Southern Paiute, Western Shoshone, and Owens Valley Paiute and Shoshone tribes.

As of 2014, the Yucca Mountain Repository has not yet accepted spent fuel rods or nuclear wastes. Various delays include lawsuits filed by indigenous nations and environmental groups, regulatory hurdles at the Environmental Protection Agency and the Nuclear Regulatory Comission (NRC) and capacity inadequacies.[20] U.S. nuclear spent fuels now are in temporary storage at each nuclear plant, on-site or near the reactors. The Nuclear Energy Institute, America's nuclear industry trade association, estimated that "20 reactors at 16 commercial nuclear sites will need additional storage space by 2010 and at least 22 more reactors will need additional storage by the end of 2015."[21] The current statutory capacity of the Yucca Mountain Repository, 70,000 tons, may prove inadequate because more than 50,000 metric tons of spent nuclear fuel is stored at more than 100 sites above the ground in 39 states, with American nuclear power reactors producing an additional 2,000 metric tons of spent fuel every year.[22]

Serious radioactive contamination has put some areas outside consideration. The lands of three North American indigenous nations have been designated as "national sacrifice zones:" the Yakima and Umatilla reservations in southern and southeastern Washington, and the Nez Perce reservation on the Idaho-Washington border.[23] However, wind and water, microbes, insects, seeds, birds, and other life-forms cannot read warning signs and pass freely from one ecological niche to another.[24] Even humans who can grasp the warnings often pay them little mind, confused or deluded by misinformation, corporate propaganda, and government restrictions—as the porous boundary of the Daiichi exclusion zones show. No clear line has been drawn between excessive radioactivity and life.

Even after nuclear wastes are removed, reactor buildings retain dangerous contamination. U.S. law requires power companies to dismantle decommissioned reactors and "clean up" contaminated sites. Corporations have been setting aside cleanup funds, but since no reactor has yet been completely dismantled, the true costs and risks of this process remain unknown.

WAFFLING AND WORSENING

To worsen matters, in Japan the slow progress of the Rokkosha plutonium breeder reactor meant insufficient fuel for nuclear power plants—leading to their use of mixed-oxide (MOX), plutonium, and uranium fuel rods—more powerful than uranium in producing heat and if accidentally aerated. This lethality prompted some to call for caution.

On July 31, 2007, Dr. Edwin S. Lyman, senior staff scientist of the Union of Concerned Scientists, appealed to Samuel L. Bodman, Secretary of Energy, U.S. Department of Energy, to delay construction of the MOX-powered Savannah River Site facility. Lyman asked Bodman "to demonstrate your commitment to the global nuclear nonproliferation regime by delaying construction of a MOX plant until 180 days after the International Atomic Energy Commission ensured that safeguards had been applied to meet the safety standards of the Global Security Program with Russia and other nations."[25]

Such concern has been brushed aside, and MOX fuel now powers plants in Iran, Japan, and the U.S. The whole and damaged Japanese nuclear power plants have released or are releasing into the atmosphere tons of jeopardy to the health and lives of thousands, perhaps hundreds of thousands, some yet unborn. WikiLeaks releases have shown that Japanese officials ignored warnings about the design of nuclear reactors years ago. They cannot safely use MOX fuel. Nor can anyone else.

MOX AND A DANGEROUS MOMENT

By March 16, 2011, with the nuclear crisis and the realization that immediate action was needed, both intensifying, the U.S. elevated the tone expressed in a "Full Committee Briefing on Nuclear Plant Crisis in Japan and Implications for the United States." For days experts batted around solemn statements and statistics: Dr. Gregory B. Jaczko, chairman, Nuclear Regulatory Commission; Mr. Anthony Pietrangelo, cr. vice president and chief nuclear officer at the Nuclear Energy Institute; Dr. Edwin Lyman, senior scientist for global security at the Union of Concerned Scientists.

Jaczko countered a Japanese denial that all water was gone in one of the spent fuel pools at the Fukushima nuclear plant. According to the Associated Press: "If NRC Chairman Gregory Jaczko is correct, this would mean there's nothing to stop the fuel rods from getting hotter and ultimately melting down."

White House spokesman Jay Carney said that the situation in Japan was "deteriorating and fast-moving" and an unarmed official was reported by ABC to have said, "They need to stop pulling out people—and step up with getting them back in the reactor to cool it. There is a recognition this is a suicide mission." Difficulties caused by evacuations were blamed for "escalating" the chances of a meltdown.

The White House advised Americans within fifty miles of the Fukushima nuclear facility to evacuate; and plant employees were temporarily forced to retreat as radiation levels "soared." And according to an official, the U.S. believed a larger evacuation zone should be imposed and that the next 24–48 hours were "critical."[26] But the Japanese government delayed enlarging the evacuation, could not stop the evaporation from nuclear fuel pools, and the talking went on.

NOTES

1. Peter Eisler, "Contaminated Water Heightens Concerns Over Nuclear Plant," *USA Today*, March 25, 2011.
2. Thomas Maugh, "Japan's Prime Minister Criticizes Nuclear Plant's Owner," *Los Angeles Times*, March 29, 2011.
3. "TEPCO Says Plutonium Found on Quake-Damaged Plant Grounds," *CNN World*, Mar. 28, 2011.
4. Ibid.
5. Ibid.
6. Hiroko Tabuchi & Ken Belson, "Contaminated Water Escaping Nuclear Plant, Japanese Regulator Warns," *New York Times*, March 28, 2011.
7. It behaves like U-235 and its fission releases a similar amount of energy. The higher the burn-up, the less fissile plutonium remains in the used fuel. Typically about 1 percent of the used fuel discharged from a reactor is plutonium, and some two thirds of this is fissile (c. 50 percent Pu-239, 15 percent Pu-241).
8. International Physicians for the Prevention of Nuclear War / The Institute for Energy and Environmental Research, Plutonium: Deadly Gold (Cambridge, MA: International Physicians Press, 1992); Julie Gordon, "Canada's Port Hope Reaps Uranium's Rewards and Risks," *Reuters U.S.*, June 2, 2011.
9. Richard Krooth, John Berger re: the Nuclear Option in Palo Alto, CA, June 5, 1976; Nuclear Information and Resource Service, "High-Level Radioactive Waste" (Nuclear Information and Resource Service, Washington, D.C., 1995).
10. Interview with John Berger; International Physicians.
11. Land and Life Film Productions, *Rokkash* or, Color (Aomori International Uranium Forum, 1991); Takagi, Jinzaburo, "Critique of Japan's Nuclear Energy Program: Collected Papers of Jinzaburo Takagi, Tokyo" (Citizens' Nuclear Information Center, 95 1994).
12. DeAngelis, Fusak, "Rokkasho: Japan's Nuclear Village," *Earth Island Journal*, Spring (1996).
13. *La Tribune*. Paris. October 6, 1996.
14. Christine Bergé, "Nuclear Debate: Superphénix Embers Under the Ashes," *Le Monde Diplomatique*, April 2011, http://www.monde-diplomatique.fr/2011/04/BERGE/20399; Collected Papers of Jinzaburo Takagi: August 1996 when the Japanese village of Maki held a historic referendum, voting 61 to 31 against the construction of a new nuclear plant in their community.
15. "Japan to Abandon Troubled Fast Breeder Reactor," *APF*, Feb. 7, 2014.
16. Koide Hiroaki et al., Saiakuno Kakushisetsu Rokkasho Saishori Kojo (2012).
17. Personal interview with Berger by Richard Krooth.
18. "Russian Announces Enormous Finds of Radioactive Waste and Nuclear Reactors in Arctic Seas," *Bellona*, August 28, 2012.
19. "Will US Dump Nuclear Waste at Sea Again?" *Christian Science Monitor*, Apr. 8, 1982.
20. See U.S. EPA Yucca Fact Sheet #1, http://www.epa.gov/radiation/docs/yucca/epafactsheet1v3.pdf (last visited Feb. 9, 2014);
21. Nuclear Energy Institute, Resources & Stats Nuclear Waste: Amount and On-Site Storage, http://www.nei.org/resourcesandstats/nuclear statistics/nuclear wasteamountsandonsitestorage/ (last visited Feb. 9, 2014).
22. Press Release, U.S. Department of Energy, Department of Energy to Send Proposed Yucca Mountain Legislation to Congress (Apr. 4, 2006), available at http://www.energy.gov/3428.htm.
23. Russell Dalton, et al., *Critical Masses: Citizens, Nuclear Weapons Production, and Enviromental Destruction in the United States and Russia* (1999).

24. S. Claire Peaslee, "Sending Signals to the Future: How to Design Long-Term Warning Markers for WIPP," *Nuclear Guardianship Forum*, 1993.

25. Dr. Edwin S. Lyman, "Letter of Appeal to Samuel L. Bodman Secretary of Energy, U.S. Department of Energy" (senior staff scientist of the Union of Concerned Scientists, July 31, 1997).

26. "US Concerned Japan Facing Situation That Could Be Deadly for Decades," *ABC News*, March 16, 2011.

Chapter Six

Hubris and Megadeath by 3 A.D. (After the Disaster)

In rare moments of introspection, the nuclear industry views its dangers through flawed, if roseate, lenses and tries to outfit the public with such implements of nearsightedness. For example, it uses a safety standard for radioactive exposure tagged "As Low As Reasonably Achievable." This is a porous and actually negative measure, since it (correctly) implies that corporations lack the technical ability to keep exposure to zero and that needs of the industry rather than health effects govern standards. Even so, what has been considered a "safe" dose has been steadily decreasing over the years.

In 1910 it was 100 rem per year.
In 1934 it was 30 rem.
In 1948 it was 15 rem.
In 1958 (and to the present) it was 5 rem for nuclear workers (the bone marrow equivalent of 1,000 chest X-rays) and 0.5 rem for the public (100 chest X-rays).[1]

These standards are highly questionable, as has been often shown. John Goffman, long an investigator and analyst of radiation effects, summed up the consensus of critiques of corporate perception promotion:

> It is not clear that 5 and 0.5 rem are safe doses. . . . This may be another case—all too frequent in the nuclear industry—of making the exposure limits fit the convenience of the industry instead of scientific and health standards. Some time ago the former Chairman of the International Commission on Radiological Protection indicated that if the worker exposure dose had been reduced from 5 to 0.5 rem per year in the early days of the industry, he doubted whether the nuclear power plants of the time could have continued operating.

> A dose that is "reasonably" or economically achievable may still cause death and disease.[2]

To put it in numbers:

Single brief exposure to radiation or containers of reactor waste can be expected to bring on the following effects:

- Genetic damage at 5 to 25 rem in a minute and half or less.
- White blood cell alteration at 50 rem exosure for 3 minutes.
- Radiation sickness at 75 to 125 rem in 5–8 minutes.
- 50 percent person die-off rate at 400 rem, and near or at 100 percent at higher rem levels and exposure for half a hour to an hour.

Government reports greatly underreported deaths from the quake-flood-reactor failures but rose periodically given public skepticism and demands for truth—it was 5,000, then 10,000, then 18,000, with at least 9,400 people missing, numbers equivalent to the deadliest single days of the bloodiest World War I battles. (17,000 Frenchmen died in the first day of the Battle of the Somme).

Three months after the Fukishima disaster, the official figures were 850,000 households without power; double that with no running water. Food and gas were scarce, and millions of people were giving up on safety and consuming tainted milk, soybeans, spinach, and tap water. The government said little about the mass flight away from ground zero and maintained only porous boundaries around the ever-widening exclusion zone, back to which filtered many tried of being trampled by a confused mass of hundreds of thousands, itself further exposed to atmospheric toxins. They, like the government, not to mention the oily corporate tongues, were overwhelmed.

THE BANZAI BUMBLE

Shaken by Godzilla, officials, just as in the movies of the lizard with death-ray eyes, called in air power, waterbombing the Fukushima I (Daiichi) nuclear power reactors whose heat mounted toward meltdown. This, of course, meant more poisoned water spill-and-spume off into land, sea, and air, and since many tons of fluid fell off-target, plant reactors and spent fuel rods went on heating and radiating. Paul Gunter of Beyond Nuclear said of the situation: "The walls of defense are falling, with the melting of the cores, the collapsing of . . . the vessels. And then, with these damaged containments, these are all open windows to the atmosphere."[3]

Some experts lauded the superior safety of U.S. reactors. But investigative journalist Karl Grossman produced a 1985 report by the National Regulatory Commission acknowledging a 50 percent chance of a severe core accident among the more than one hundred nuclear power plants in the United States over a twenty-year period. These plants and waste storage areas are in every part of the U.S.; many were designed just like the Japanese plants to boil water.

WILL THE HELP COME SOON ENOUGH?

In Japan discordant views emerged within days of the initial failures of structure, power, and external energy sources. Steven Leeper of the Peace Culture Foundation, which manages the Hiroshima Peace Memorial Museum in Hiroshima, looked for the silver lining:

> In Hiroshima, we are pretty sensitive to radiation issues, and we're very sensitive to disaster issues. . . . We have a team of doctors. They left yesterday to go up into that area with their equipment to try to figure out what kind of radiation is up there. We've also prepared a lot of apartments, and the hospitals are making preparations to receive radiation victims.[4]

This we-can-handle-it response fit with Japanese reputation for resiliency, but suggests certain shortcomings in data and specifics.

In Yokohama, Keibo Oiwa, the founder of the Sloth Club, Japan's leading "Slow Life" environmental group, and a leading Japanese cultural anthropologist and environmentalist, made a more radical appraisal, one that the oligarchs would labor rightily to neutralize: "I'm realizing again that democracy is so hollow now. We do not have power. We have been controlled by the government and the Tokyo Electric Company, a private company. . . . We have to really look for a lifestyle and a way of thinking again, to live again with harmony, in harmony with nature."[5]

The government of course first said nothing dangerous had occurred and that the situation was under control. Schools in the contaminated zones soon opened, with students, those left in the area, warned to stay inside.

COST ACCOUNTING

Fatal danger aside, nuclear facilities are bad investments.

Official calculations of the cost of nuclear energy consider only the direct costs of building and operating reactors, plus mining, processing, and transporting fuel.[6] They do not take into account indirect costs to society from environmental and health damage, or the costs of accidents, cleanup, nuclear waste storage, and decommissioning. These can run up to millions of dollars.

Nuclear energy is also a tragically misdirected investment. It not only pollutes the environment and destabilizes security, it also misuses precious financial resources as well as scientific and engineering expertise. Why, then, are new nuclear reactors being promoted and built, producing ever more toxic plutonium and radioactive wastes? One answer is profit-seeking—fueled in part by government subsidies and bank loans to nuclear corporations, keeping the deadly industry alive.

Furthermore, nations which use radioactive materials to produce energy gain access to the global club of "borderline" nuclear weapons states. As long as the five permanent seats of the United Nations Security Council are assigned to the five declared nuclear weapons states, the assumption that "access to nuclear weapons equals global power" will encourage nations to pursue nuclear energy—and thus the potential of nuclear weapons—to increase their influence in global affairs.[7]

Even before the explosion on March 12th, Japanese officials said they had detected radioactive cesium-137, which is created when uranium fuel is split, an indication that much of the nuclear fuel in the reactor was already damaged.

During the meltdown the Japanese Nuclear and Industrial safety agency reported many people might have been exposed to radiation around the first plant; and Japanese news media said three workers at the facility were suffering from full-on radiation sickness, meaning they would shortly die.

Public broadcaster NHK flashed instructions to evacuees: close doors and windows; place a wet towel over the nose and mouth; cover up as much as possible. At a press conference, Mr. Edano called for calm. "If measures can be taken, we will be able to ensure the safety of the reactor," he insisted.[8]

He was knowingly lying. There was no way the nearby population could protect itself, short of complete evacuation.

The blast March 12th afternoon destroyed part of the facility, and officials said that they had filled it with seawater to *prevent full meltdown* of the nuclear fuel. That step would only be taken in extreme circumstances because ocean water would likely permanently damage the aged plant. How much damage the fuel suffered remained uncertain at that point, though safety officials insisted repeatedly through the day that radiation leaks outside the plant remained small and did not pose a major health risk—another straying from the truth to stop public panic.

However, they also told the International Atomic Energy Agency that they were preparing to distribute iodine pills, which supplements and helps protect the thyroid gland from radiation exposure, to people still living near the Fukushima I (Daiichi) plant and a second nuclear plant, Daini, or Fukushima II, that suffered damage in the quake, about ten miles away. True or false, help came too late—the population should have been immediately moved in all directions at least sixty miles away from the Daiichi plant. U.S. citizens working in Japan were moved fifty miles away, with the tragic exception of those involved in rescue and relief operations.

NOTES

1. John Goffman, Nuclear scientist, n.d.
2. Goffman, supra note 1.
3. Paul Gunter, Beyond Nuclear, n.d.
4. Steven Leeper (Hiroshima: Peace Culture Foundation, Hiroshima Peace Memorial Museum, n.d.).
5. Keibo Oiwa (Yokohama, Japan: Sloth Club, "Slow Life" Environmental Group, n.d.).
6. Charles Komanoff and Cora Roelofs, *Fiscal Fission: The Economic Failure of Nuclear Power* (Washington, DC: Greenpeace, 1992).
7. Berger, *Nuclear Power*; Leonard S. Spector and Jacqueline R. Smith, *Nuclear Threshold: The Global Spread of Nuclear Weapons, 1990–1991* (Boulder, CO: Westview Press, 1992).
8. Chief Cabinet Secretary Yukio Edano, n.d.

III

In the Maelstrom

With apparently uncontrollable radiation soaring into the atmosphere, putting bad seed into the soil, and spiking the sea, Japan's oligarchy strained the limits of population manipulation and control. Evidence of distrust of government, media, and corporations was found in all layers of society. It had been alerted that bureaucracy and kowtowing to traditional methods solved no immediate problems; and costs in the future would not trickle but cascade down to the middle and lower ranks of society.

The elaborately constructed government-corporate structures offered no refuge from nuclear bullets riddling the environment; the drought of safe foods and water; the ebbing of employment and even sorely needed basic charitable help. All of Japan's exports—foods, metals, synthetics, even cars—were suspect, spurned by even those in Japan able to buy them. Outcries of pain, fear, and anger at authorities and apologists arose: the marshals of public trust were not only fallible but apparently unconcerned with truth and justice or the welfare, the very safety, of the population.

Public mistrust checked the elite's denial, diminishment, sympathy, pain-sharing, and even contrition gambits. Tokyo Electric's chairman resigned but his successor was another insider. New political parties offered half-measures. The center could not hold, especially when people realized Tokyo elites stood to profit from the bear market in confidence as they had in rosier times.

It seemed the centralization of Japan during the Meiji Restoration and the post–World War II state and industrial megaplex (*zaibatsu*) restoration under America's Supreme Command was in accelerating reversal. A new Japan might emerge, more democratic and equitable perhaps, certainly more compassionate. Or a new, German-style, post–World War I, militarized Weimar Republic under press lords and an alliance of the governing state with nuclear power companies.

FALSITY VS. FALLOUT

The last hopes of the public were raised by government promises in mid-April that the Daiichi burial would be completed and normal operations resumed nearby by the close of 2011. Tokyo Electric at first aimed to reduce radiation leaks in three months and to cool the reactors within nine months. Tsunehisa Katsumata, the chairman of Tokyo Electric, dumped in May, spoke of "cold shutdown" allowing the tens of thousands of families evacuated from the area around the facility to return home as soon as possible. He said the firm was preparing burial of the reactor building. But immediate radiation levels in the sea near Reactor Two spoke of increased dangers, rising from April 15th to 6,500 times the legal limit, up from 1,100 times a day earlier—raising fears of fresh radiation leaks and purposeful dumping of fissionable seawater. Even the government showed signs of suspicion about Tokyo Electric's pledges and would in May set off a sensation by suggesting it would not bail out the bank that had bailed out the corporation.

But through April, the nuclear operators insisted that after radioactive elements had been flooded in a "cold shutdown," they could do a quick cleanup, encase damaged reactor buildings, and clean up radiant soil, fuel, and wastes. They actually played video games, deploying robots from British defense contractor QinetiQ, operated from a standard video games console, to enter the reactor and gauge radiation and temperature. These drones were supposed also to bash and pile dirty pipes, tanks, and walls, clear the rubble, and retest for radiation. Japan, though its droids had set world standards, has none adapted for sopping up after meltdowns.[1]

The following repeats the "Introduction" covering the actual time required for nuclear cleanup, that slowly became known to the public after May 1, 2011, when two conglomerates submitted proposals for the billion-dollar job: the quickest it could be done was ten years, a decade that would include nothing about leaked reactor plutonium, uranium, or MOX fuel already abroad.

It would involve tear-down, recovery, and "disposal" of used fuel in safe storage, covering the nuclear plant in sand, boric acid, and cement, cleansing the soil, and setting up and maintaining a police perimeter at least sixty miles away. When safe levels of radiation were achieved—the nuclear agency and the government to decide what "safe" meant—then life could return to "normal," also defined by the usual sources. Toshiba said it could reach this goal in a decade. Toshiba's proposal team that said it could stabilize the one hundred square miles around Fukushima included engineers from Westinghouse—in which Toshiba has majority stock holdings—and Babcock & Wilcox, experienced in hazardous waste disposal. It and Westinghouse had helped shut down the blown Three Mile Island reactor.

Hitachi issued a less glowing but more transparent estimate that posited a three-decade effort to return the site to a "green field" state. Hitachi, the country's second biggest supplier of reactors, had fifty experts working on its dismantling plan, some from Daiichi architect General Electric, others from the American nuclear operator Exelon and the giant engineering company Bechtel.

Referring to Toshiba, Hitachi spokesman Yuicgi Izumisawa said. "At Hitachi, we are baffled by what technology would allow completion in ten years." Hitachi predicted 10 years would only be time enough to get things started by extracting the nuclear fuel rods and putting them in well-padded coffins for transport to a nuclear mausoleum. Only then could the dismantling of the plant's structures begin, followed by the cleanup of the remaining radiation. "You basically need to dismantle the plant from the inside, and the inside is still very radioactive." Both cleanup bidders would face possibly unforeseen production disruptions that delayed or cut off supplies and labor.

Comparisons are camp, and when it comes to nuclear reactor cleanup, they mean little. Chernobyl still simmers, though its reactor is a ruin under sand, lead, and concrete; wastes and still-reactive fuel keep an area totally uninhabitable; the cleanup of the 1986 Pennsylvania Three Mile Island blowout took fourteen years, cost some $1 billion, though only one reactor suffered a partial meltdown of the nuclear fuel rods. The two surviving reactors are to be torn down in 2014, with nowhere on Earth to store safely their used plutonium rods.

Tetsuo Matsumoto, a professor of nuclear engineering at Tokyo City University, said how long the decommissioning process would take depended heavily on the state of the nuclear fuel, as yet seen only by robots: "Will it still be shaped like rods? Or [like Chernobyl] will it have melted and collapsed into a big mass?" he said. "It could be 10 years or 30. You just won't know until you open up the reactor."[2]

It's going to be a long wait, if Ukraine's Chernobyl reactor No. 4 is any example. It blew out on April 26, 1986, and isotopes keep leaking out today, draped in a shapeless steel sarcophagus. The Ukraine government is still struggling to contain the radiation, building a two-part shield assembled on rails, which the heavily protected, six-hundred-man crew—with the help of powerful hydraulic jacks—will later use to move the structure toward and finally over the reactor to isolate it. The delicate task of moving the shield three hundred meters is expected to take four days. Once in place, the shield will be monitored from afar to avoid any human activity close to it. It will be equipped with a ventilation system and sensors to evaluate corrosion levels—

a virtual mausoleum where humankind can offer condolences and hopes that none like this should ever again appear in its backyard.

NOTES

1. Roland Buerk, "Japan Nuclear Crisis 'Over in Nine Months'", BBC, 17 April 2011.
2. See Hiroko Tabuchi, "Nuclear Cleanup Plans Hinge on Unknowns," April 16, 2011.

Chapter Seven

Fail-Safe Not Foolproof

Risky business fail-safe plans often miss the mark, especially in energy and chemical entreprises where fires, spills, and explosions are almost commonplace. Not so long ago, one might see in front of refineries or other factories billboards proudly announcing the number of days since the last fatal accident—usually not in the three digits. To choose a more recent example than say, the blowing away of Texas City in the late forties, West, Texas in 2013, or the Exxon Valdez spill, BP (Beyond Petroleum, as it advertises itself) asked U.S. regulators to let it resume drilling in the Gulf of Mexico, [*New York Times* website April 3, 2011], aided in its quest by the state of Texas which kept the Environmental Protection Agency busy with a lawsuit aimed at relaxing pollution standards.

Rebel advances in the Middle East, threatening oil shipment, also help. BP's move came a year after a rig that the British oil company leased exploded. According to the newspaper, BP in 2013 asked permission to continue drilling at ten existing deep water production and development wells in the region in exchange for embracing, what BP put forth as stricter safer and supervisory rules. The blast, the *Deepwater Horizon* explosion in April 2010, spilled more than 200 million gallons of oil from BP's well a mile undersea, according to government estimates that BP unsurprisingly disputed. Ridley turtles, which once laid their eggs on Texas sands, have not been seen since.

Fail-safe plans mean about as little in nuclear operations. The Japanese national news agency, Kyodo, for example, reported radiation of up to nine times the normal level was briefly detected in Kanagawa prefecture, south of Tokyo, far from the exclusion zone in Fukushma Prefecture. The report came after people had been displaced from the area to what was termed safety and the situation was under control, according to the solemn faces of television.

Further radiation rises have been reported in many areas in the northern Pacific basin and to the east of Japan.

In fact the explosions following the tsunami and power failures cmpletely breached protection around the power units at Fukushima, once promoted as impregnable by men clothed in suits and titles that connoted expertise. As with BP and others, after-the-fact corporate statements denied the extent of the damage, cracked substructures, burst steampipes, fires, and the rest; but unlike developments in the BP case, the government openly seconded the BP claim that aftereffects were more an issue of psychology than biology. As with Chernobyl, the numbers are not in and may never be, but a few scientific facts are now more than Wikileaks general estimates. A primary question might be how much punishment human cells can take.

The World Health Organization (WHO), range for exposure of the average person's to radioactive material is 3mSv in a year. But the International Atomic Energy Agency (IAEA) reports: "Dose rates of up to 400 millisievert per hour (mSv/hr) have been reported at the Dai-ichi site." The U.S. National Institutes of Health offered its condolences: a dose approaching 1,000 millisieverts (1,000mSv) would certainly induce radiation sickness. But luckily or not in the 1986 Chernobyl disaster, some workers who received doses above this level are still alive, generally suffering pain and multiple cancers.[1]

The damage control effort, psychological department, came up with an interesting not-in-my-backyard ploy. Japan's meteorological agency reported that the prevailing wind in Fukushima in spring of 2011 was eastward into the Pacific and would help carry away any radiation.[2] The agency was not quite on the button, if one can believe many reports that those fleeing from the disaster to the higher ground in western Fukushima and elsewhere ran before a stff offshore east-to-west breeze. Not to mention the usual NIMBY shortsightedness: the radiation carried away by wind and water endangered others' backyards, food supplies, health, and the genes they might pass on to the future. At any rate, reports of some 20,000 deaths and dangers for maybe five times that many around Fukushima Prefecture was small potatoes compared to much gorier episodes in the century-long slaughterhouse just past. And the experts simplified for the public the matter of reactor repair.

It seemed at first something of a plumbing problem. Certainly the radioactive water leaks had caused three meltdowns, blown the top off at least one reactor building, and aerated enough nuclear toxins to put ten million people in permanent jeopardy, but from the proper perspective the problem could be seen to resemble a water heater malfuncton. One had to turn off power, let things cool, mop up and dispose of puddles, check and possibly repair the piping and controls, and then get the juice flowing again. The problems at the reactor, however, were complicated by the toxicity of the water and the difficulty of turning the power off, absent internal or external long-lasting power. Its shortage even forced emergency staffers at the reactors out to parking lots to cannibalize car and bus batteries as DC sources.

In its mop-up, or removal of toxic water under the turbine buildings of reactors 1, 2 and 3, Tokyo Electric needed to siphon off a nearly full ditch of radioactive water. Pumps ran for several hours on March 31st to lower the water level under Unit 1 and transfer it to a storage tank. Unit 2's drain trench was too radioactive to approach, topping more than 1,000 millisieverts per hour, but efforts continued, slowly, to dry up the turbine trenches of all four reactors, where the water depth ranged from 20 cm to 1.5 meters. The pump-out was complicated by the continuing reactivity of the nuclear fuels exposed in the reactors and the emanations from the contaminated water. Reactor drippings had eaten through steel and concrete, the water in some places was hot enough to separate into flammable and explosive hydrogen and oxygen, and a mere approach to the workings of the reactor/boiler could wound or kill—no matter what the suits said, it was not your average boiler blowout.

The money men did, however, get pushed a little beyond their comfort zones, even into the relam of creativity, with fantasies of a protective ice wall or a synthetic resin shroud for Daiichi, or even schemes for recycling nuclear wastes to produce more nuclear power and wastes and on . . .

NOT ALL OVER: THE BLIND SPOT WIDENS

Meanwhile, the detections of danger grew. On March 30th, the Geneva-based International Atomic Energy Agency identified 2 million becquerels of radioactive substances per square meter in soil samples from the village of Iitate twenty-five miles (forty km) northwest of the Fukushima No. 1 power plant. This was double the threshold at which the IAEA itself would order an evacuation. The IAEA urged Japan to expand the current evacuation zone of twelve miles (twenty km) around the plant. But the government on March 31st waffled, merely advising residents within twelve to fifteen miles (20–30 km) of the plant to stay indoors. Secretary Yukio Edano only said the government might consider expanding the mandatory evacuation zone if the higher levels of radiation continued. "If a person is exposed to the radiation levels over the IAEA criteria for a long time, it might affect their health because radioactive substances could accumulate in the body," Edano said. "If that is the case, we need to consider evacuating residents from the area." But at the end of the month of the meltdown, he said the government would conduct air and soil monitoring in a more detailed manner at Iitate, where about one hundred people still hung on to their homes.[3]

These few would be helplessly irradiated, as the government's indecision lingered into April and the reactors continued to blow irradiated steam, leaked water, and ionized the soil and ocean. The unit-by-unit score then was as follows:

- Reactor 1: Damage to the core from cooling problems. Building punctured by gas explosion. Radioactive water detected in reactor, basement, and groundwater.
- Reactor 2: Damage to the core from cooling problems. Building punctured by gas blast; containment damage suspected. Highly radioactive water detected in reactor and adjoining tunnel.
- Reactor 3: Damage to the core from cooling problems. Building punctured by gas blast; containment damage possible. Spent fuel pond partly refilled with water after running low. Radioactive water detected in reactor and basement.
- Reactor 4: Reactor shut down prior to quake. Fires and explosion in spent fuel pond; water level partly restored.
- Reactors 5 & 6: Reactors shut down. Temperature of spent fuel pools surged then lowered.

At the same time, Japan's Nuclear and Industrial Safety Agency (NISA) announced that the level of radioactive iodine-131 in seawater near the plant was 4,385 times the maximum tolerable amount, the highest reading since the crisis began. It revealed that other highly radioactive materials had been detected, including cesium-134 at 783.7 times the maximum amount permitted, and cesium-137 at 527.4 times the legal limit. The radiation lethal levels were well-known, though: The half-life of cesium-137, or the time its radioactivity dissipates by half, is thirty years, compared with eight days for iodine-131 and two years for Cesium-134.

Not to worry. NISA repeated that such levels posed no immediate health risk: "Although the figure is on the rise, it will not immediately affect (the health of) residents nearby as the evacuation area is set at a 20-km radius from the nuclear plant and fishermen are not working in the area," said agency official Hidehiko Nishiyama. "Radioactive material will spread and dilute by the time people take it in through [ingestion of] marine products."[4]

Nishiyama simply advised that NISA would increase the number of ocean monitoring sites. The science ministry already had been monitoring radiation about thirty kilometers offshore, while NISA would continue to conduct checks about fifteen kilometers from the coast. Nothing was done to immediately evacuate the hundreds in the radiation zones.[5]

LONG-TERM EVACUATION PLANS—IS IT FOREVER?

Would evacuation of the more fortunate be forever? Secretary Edano and Prime Minister Naoto Kan hedged, as temporary shelters overflowed with many of the 70,000 evacuees and 136,000 people who lived within twenty to thirty kilometers of the plant were encouraged by the authorities to leave or to stay indoors. In a televised address, Prime Minister Naoto Kan also hedged—the government was "prepared for a long-term battle" at the plant—one, he said, that would be won. "At the current stage, we cannot say that the plant has been sufficiently stabilized. But we are preparing for all kinds of situations and I am convinced that the plant can be stabilized. We cannot say at this stage say by when this will happen, but we are trying our best," he postulated.

The authorities continued to lie, resisting calls from the UN's atomic agency to expand the exclusion zone around the plant, even after highly radioactive water continued to leak from the reactor 1, found in groundwater fifteen meters below it. Eventually it would affect drinking water as concentrations enlarged, and radioactive material continuously leaked from the damaged plant, feared Japan's own nuclear and industrial safety agency (NISA).

SHIELDING THAT WORKS NOT AT ALL

By April 1st, the radiation, which had halted shipments of certain vegetables, dairy produce, and other foodstuffs from four nearby prefectures, had been widened to include beef, the health ministry said. Yet the government continued to insist that no water or food contamination had reached levels that would be harmful to people's health.[6] And Iodine-131 was everywhere in the region, with pregnant women already affected; and their I-131-impacted breast milk liable to be transferred to new-born infants. The Japanese milk supply was also permeated by iodine from cows and goats grazing on pastureland. Water too was suffused with contaminate I-131.

U.S. petitioners immediately asked the American maker of no-water-required formula milk to send it to Japanese mothers nursing their babies and where there was insufficient baby formula that mothers could use without water. Sending this type of milk, such as Similac baby formula, would assuage the swelling worry about food and water supplies in the devastated area and other prefectures. Protests slowly emerged, but more than 10,000 people showed up for two antinuclear protests in Tokyo on April 11, some with signs and shirts emblazoned "No nukes!"

AND THE RADIATION GOES ON

Radiation will not now or perhaps ever completely subside at Daiichi or throughout Fukushima. It will remain an invisible turning point in history—the point at which devastating fission irradiated the environment and inhabitants. No memorial stones will mark it; no company or government officer can ever offer acceptable apologies or contrition for the devastation. The publicity about it tends to distort its real meaning, to relegate it to a footnote in history; even as robots and cameras sent into the plants detect more dangers Prime Minister Abe announces a three-point economic agenda, first of which is to lower corporate taxes.

The second aim in Abonomics is to use more robots in industry, oseteniby to meet the problem of Japan's declining workforce, but perhaps a nod to the fact that the blast site remains unsafe for humans. In summer 2011, radiation levels near the plant's No. 1 reactor reached 4,000 millisieverts per hour—an exposure level roughly equivalent to 40,000 chest X-rays. Even the Tokyo Electric Power Company said it would not send workers into such a line of fire. And workforce decline could be due, more than infertility, to the doubt and worry of those—especially in Japan's northeast—exposed to aerated nuclides, a poisoned sea, and an estimated 40,000 tons of highly contaminated radioactive water overflowing from the reactors[7]

NOTES

1. Richard Black, "Radiation Falls at Japanese Plant," BBC, March 15, 2011.
2. Eric Talmadge and Mari Yamaguchi, Tokyo, Associated Press, March 15, 2011.
3. Chief Cabinet Secretary Yukio Edano., n.d.
4. Hidehiko Nishiyama, NISA official, Japan, n.d.
5. Kanako Takahara and Kazuaki Nagata, "High Radiation Found Outside No-go Zone; but Edano Says Evacuation Area Won't Be Expanded for Time Being," *Japan Times*, April 1, 2011.
6. BBC, April 1, 2011, News edition.
7. *Democracy Now*, June 6, 2011.

Chapter Eight

From Sea to Shining Sea: Radiation without End

Over forty years ago the National Academy of Science in *The Effects on Populations of Exposure to Low Levels of Ionizing Radiation*, reported that carcinogenesis could result from radiation of only one or a few body cells. They based the finding on extrapolation from data from the atomic bomb survivors of Hiroshima and Nagasaki, patients irradiated therapeutically, and those occupationally exposed. In 2006 the Academy in another report weighed radiation effects in the low-dose range of <100 mGy (10 rad or rem—a quick single X-ray), with emphasis on the lowest doses. Effects that may occur as the radiation is delivered chronically over several months to a lifetime were thought to be most relevant in nuclear disaster survivors or those indirectly exposed to radiation.

At low doses, damage is caused by the passage of single particles that can produce multiple, locally damaged sites leading to DNA double-strand breaks (DSBs). DNA DSBs in the low-dose range can be quantified by a number of novel techniques, including immunofluorescence, comet assay, chromosome aberrations, translocation, premature chromosome condensation, and others. Some of these indicators of DSBs show linearity down to doses of 5 to 10 mGy. Most significantly, a dose-response relationship with mortality from nonneoplastic disease was demonstrated in 1992, and subsequent analyses in 1999 and 2003 have strengthened the evidence for this association. Statistically significant associations were seen for the categories of heart disease, stroke, and diseases of the digestive, respiratory, and hematopoietic systems regardless of the type of dose-response: linear, pure quadratic, or with a threshold as high as 0.5 Sv.[1]

The National Academy analysis of low-level effects may turn out to be optimistic or simplistic, given the Fukushima disaster's spewing I-131 and other lethal nuclides over concentrated populations in the danger zone and the government's slow set up and moving people from the disaster zone. Authorities equivocated after the earthquake, tsunami, nuclear explosions, and massive irradiation and finally left survival largely in the hands of the traumatized and weak, impoverished and helpless. No special care was taken to guard pregnant or nursing mothers or children.

The U.S. Environmental Protection Agency has made public only a bit of nuclear chemistry:

> Iodine (chemical symbol I) is a nonmetallic solid element. There are both radioactive and non-radioactive changed atoms of iodine. Of the radioactive isotopes Iodine-129 and -131 are the most important. Some isotopes of iodine, such as I-123 and I-124 are used in medical imaging and treatment, but are generally not a problem in the environment because they have short half-lives.
>
> Cesium (Cs) is a metal that may be stable (nonradioactive) or unstable (radioactive). The most common radioactive form is Cs-137. Another common radioisotope is Cs-134. Cesium-137 is much more significant as an environmental contaminant than cesium-134. It is also very useful in industry for its strong radioactivity.[2]

FIDDLING AND BURNING

The government, Tokyo Electric, and state-regulated media conducted a charade of concern and hope that still goes on as they misinform people about thin and erratic emergency efforts and the depth and breadth of the radioactivity. For example, as the government merely cautioned citizens to stay indoors, seawater taken on April 3, 1,080 feet away from the Fukushima Daiichi plant revealed levels of iodine-131 over 4,000 times normal, and levels of cesium-137 some 527 times normal, and heightened radiation was reported all the way across the Pacific.

The measures mandated—duck-and-cover-type self-help advice for families and using sawdust, shredded newspaper and an absorbent powder to stop the leak of radioactive water into the sea—were hardly credible: "I've never heard of anything like it at a nuclear power plant," said Itsuro Kimura, emeritus professor at Kyoto University and director of the Japan-based Institute of Nuclear Technology. What was needed, he emphasized, was to get cooling systems back online at the plant's six reactors, because the only these systems, properly functioning, could reduce nuclear heat: "For now, they must stop leaks any way they can."[3]

A few days later, samples of water used to cool reactor No. 2 showed 5 million times the legal limit of radioactivity, adding to fears that contamination was spiking and spreading far beyond the disaster zone. The government held a half-hearted debate about imposing radioactivity restrictions on seafood but delayed action. Repair workers, weeks after the disaster, began a suicidal struggle in hellish surroundings to restart the cooling pumps. Until and if the pumps came back, seawater would be poured over the reactor cores, corroding pipes and joints and washing wastes and radiated liquid out to the ocean and spilling on the ground as crews tried to get core fuel and water into transports for storage in places of varying degrees of readiness and capability for the influx. The 60,000 tons—some 15 million gallons—of poison filled overtaxed shipment and storage facilities Tokyo Electric relabeled "low-level radioactive seawater" and began pouring 11,500 tons—almost three million gallons—into the Pacific.[4] As this hot gloop hit the sea, its plumes and fogs would later rain radioactive fissions of iodine-131, cesium-137, and cesium-134 wherever clouds and mists traveled.

RISKS REAL FOR CESIUM-137

People may ingest cesium-137 with food and water, or may inhale it as dust. If cesium-137 enters the body, it is distributed fairly uniformly throughout the body's soft tissues, resulting in exposure of those tissues. Slightly higher concentrations of the metal are found in muscle, while slightly lower concentrations are found in bone and fat. Compared to some other radionuclides, cesium-137 remains in the body for a relatively short time. It is eliminated through the urine. But repeated or constant exposure may dangerously impact health since effect of chronic radiation can be cumulative if repeatedly exposed.

Exposure to radiation from cesium-137 results in increased risk of cancer. Everyone is exposed to very small amounts of cesium-137 in soil and water as a result of atmospheric fallout. Exposure to waste materials from contaminated sites or from nuclear accidents can result in cancer risks much higher than typical environmental exposures.

If exposures are very high, serious burns, and even death, can result. Instances of such exposure, beyond background levels, are rare. An example of a high-exposure situation would be the mishandling of a strong industrial cesium-137 source. The magnitude of the health risk depends on exposure conditions: strength of the source, length of exposure, distance from the source, and whether there was shielding, such as metal plating, between receiver and source.

Several different medical tests can determine exposure to cesium-137, though routinely unavailable in a doctor's office because they require special laboratory equipment. Some tests can measure the amount of radionuclides in urine, or in fecal samples, even at very low levels. A technique called "whole-body counting" can detect gamma radiation emitted by cesium-137 in the body. A variety of portable instruments can directly measure cesium-137 on the skin or hair. Other techniques include directly measuring the level of cesium-137 in soft tissues samples from organs or from blood, bones, and milk.

Cesium-137 that is dispersed in the environment, like that from atmospheric testing, is impossible to avoid. However the exposure from cesium-137 in the environment is very small. Serious exposure is unlikely. People most likely to accidentally encounter a cesium-137 source typically work in scrap metal sorting, sales, and brokerage; metal melting and casting; and in municipal landfill operations. They may unwittingly encounter an industrial instrument containing a sealed cesium-137 radiation source.

Can Japan's Tokyo Electric and the Japanese Nuclear Protection Agency equal what the EPA is doing about cesium-137? No reasons exist to prevent or impede them.

Both EPA and the Nuclear Regulatory Commission regulate cesium-137. The Nuclear Regulatory Commission licenses its use. EPA has several regulations that protect us from cesium-137 in the environment. These include standards for the maximum amount of cesium-137 that nuclear facilities may release to the air, and maximum levels for cesium-137 in drinking water. EPA also sets risk-based criteria for cleanup of soil and groundwater at sites contaminated with cesium-137 that must be met before the site can be approved for public use.[5]

IODINE-129 AND IODINE-131 IN THE ENVIRONMENT

If, then, this is not reason enough to be out of the reach of radiation exposure, consider trying to evade two types of invisible fissions of iodine. Iodine-129 and iodine-131 are gaseous fission products that form within fuel rods as they give off energy.

Unless reactor chemistry is carefully controlled, reactor fuel rods can quickly increase pressure and corrode the rods. As the rods age, they may develop cracks and holes. Breached release of radioactive iodine in the cooling water around the rods circulates throughout the system, ending up in the airborne, liquid, and solid wastes from the reactor. From time to time, reactor gas-capture systems release gases, including iodine, to the environment—monitored, ideally, under strict government regulations.

Anywhere *spent nuclear fuel* is handled, iodine-129 and iodine-131 may escape into the environment. Reprocessing this remnant to recover plutonium and other valuable materials releases more iodine-129 and -131 into the airborne, liquid, and solid waste processing systems.

Because of concerns about nuclear weapons proliferation and terrorist acts, the U.S. no longer reprocesses spent fuel. It now resides in temporary storage at nuclear power plants across the country. If the nuclear waste repository at Yucca Mountain opens, it will provide "permanent" disposal for spent nuclear fuel and other high-level radioactive wastes and the short-lived iodine-131 it contains will decay quickly and completely. However, the *long-lived iodine-129 will remain active for millions of years*. Keeping it from leaking into the environment requires carefully designed, long-term safeguards.

The *detonation of nuclear weapons* also releases iodine-129 into the environment. Atmospheric testing in the 1950s and 1960s released radioactive iodine to the atmosphere which has been disseminated around the world and is now found at very low levels in the environment. Most I-129 in the environment came from weapons testing. Under the right conditions, radioactive iodine can disperse rapidly in air and water but it combines easily with organic materials in soil. This is known as "organic fixation" and slows iodine movement in the environment. Some soil minerals also attach to, or adsorb, iodine, which also slows its movement. Health consequences are yet to be fully investigated.

Radioactive iodine can be inhaled as a gas or ingested in food or water. It dissolves in water so it moves easily from the atmosphere into humans and other living organisms and can also be transmitted via body fluids. Nuclear plant emissions in addition to past nuclear tests increase the level of general exposure. I-129 and -131 can settle on grass where cows can eat it and pass it to humans through their milk. It may settle on leafy vegetables and be ingested by humans. Iodine isotopes also concentrate in food fish.

Also, doctors may give thyroid patients radioactive iodine, usually iodine-131, to treat or help diagnose certain thyroid problems. The tendency of iodine to collect in the thyroid makes it very useful for highlighting parts of its structure in diagnostic images and as well a prime target for accumulation of radiation.

In the body, iodine has a biological half-life of about one hundred days but different radiation half-lives for various organs: thyroid—one hundred days, bone—fourteen days, and kidney, spleen, and reproductive organs—seven days. Since it concentrates in the thyroid gland, a radio assay of the thyroid can determine the level of exposure to many of its isotopes. However, I-129 has low activity and emits extremely low energy beta particles, making a radio assay problematic. I-131 assays, on the other hand, are available through most major medical centers.

The thyroid has a limited capacity and cannot distinguish radioactive from stable iodine, so it absorbs radioactive iodine in whatever mix it meets. If a nuclear accident releases much radioactive iodine, agencies may distribute doses of stable iodine to block absorption of dangerous iodine by the populace's thyroid glands. So stable iodine can be a guard but also a health hazard in long-term use. Iodized table salt normally suffices to maintain health.

The U.S. EPA has limited the release of radionuclides, including I-129 and I-131, to the environment. The limits cover airborne and liquid releases from nuclear reactors, airborne emissions from industrial and governmental facilities, and from radioactive waste disposal systems. Air and sea emissions, of course, travel farther faster than polluted dirt, and the sea is an especially dangerous vector with its dispersals slower than those in the air and its occupants capable of absorbing nucleides.

The pre-Fukushima sea was already something of a toxic dump, as the famous French oceanographer Jacques Cousteau warned:

> During the past thirty years, I have observed and studied the oceans closely, and with my own eyes I have seen them sickened. Certain reefs that teemed with fish only ten years ago are now lifeless. The ocean bottom has been reaped by trawlers. Priceless wetlands have been destroyed by landfill. And everywhere are sticky globs of oil, plastic refuse, and unseemly clouds of poisonous effluents. Often, when I describe the symptoms of the oceans' sickness, I hear remarks like "they're only fish" or "they're only whales" or "they're only birds." But I assure you that our destinies are linked with theirs in the most profound and fundamental manner. For if the oceans should die—by which I mean that all life in the sea would eventually cease—this would signal the end not only for marine life but for all other animals and plants on earth, including man.[6]

Save what whales?

SORCERY'S APPRENTICES

After the meltdown, U.S. engineers on-site offered advice—with the fate usual to most advice. The U.S. experts said a wide array of dangers could arise and remain. The very measures taken to stabilize the plant could create more problems than solutions. They used data provided to the team by numerous Japanese and American organizations, including the electric power company, the Japan Atomic Industrial Forum, the United States Department of Energy, General Electric, and the Electric Power Research Institute, an independent, nonprofit group.[7] Each week disclosed new crises. Nuclear engineer David A. Lochbaum, experienced with the kinds of General Electric reactors used in Japan, said conditions at three separate reactors made a successful repair or even stabilization uncertain. (Lochbaum now directs the nuclear safety project at the Union of Concerned Scientists and was the lead author of that organization's study, *Fukushima: The Story of a Nuclear Disaster*.)

Each week after March 11, new troubles arose, with wall and pipe rupture, explosions, and corroding connections and machinery topping the list. Stress from wavelets and weight of radioactive cooling water warped containment structures. The seawater salt and fuel rod melt interfered with passage of used water, so it heated enough to break into hydrogen and oxygen atoms as it hit reactor hot spots. The plant meanwhile leaked plutonium-laced fluid into the soil *below the reactor*—though managers claimed leaks had been sealed with a liquid glass coagulant, the effect of which, of course, could not be determined by firsthand, robot, or photo observation.

Doomed workers in the house of shards scrabbled after leaks and clogged or damaged piping and hoses for seawater. They ransacked parked vehicles outside the plant for batteries to power cleaning and pollution measuring. Some of them found exit doors locked. Their heroism earned them radiation burns from new leakages and radioactive pools near the core of the reactors. The situation, aside from its chaos and clatter, could bring to mind the Dutch finger-in-the-dike tale: but here the dike was made of Swiss cheese.

In other words, a total meltdown and more explosions in all six reactors loomed. The flood-and-cool plan showed meager and dilatory results, hydrogen accumulated, and a tangled mass of radioactive trash littered the site. From time to time, hydrogen explosions rattled the plant structures, shooting out nuclear fragments, damaging the super-radioactive spent fuel pools which GE design had unfortunately placed atop the reactors. Blown as high as one mile above the nuclear units, pool water and pieces of highly radioactive material dropped like black rain over a wide area. Near Daiichi wastes, water, and fuel were haphazardly covered by workers and bulldozers, both absorbing radiation.

The U.S. Nuclear Regulatory Commission advised steps which apparently seemed like sci-fi to the Japanese. Engineers should add more boron to cooling water to prevent fuel cores from restarting nuclear reactions leading to meltdown. They should inject inert nitrogen gas into the containment structures to try to purge their explosive hydrogen and oxygen. And maybe, though the Commission did not say this, they should cross their fingers against aftershocks from the earthquake which could bring down on their heads the whole complex, groaning as it was under the overload of radioactive cooling water and pools, atop the remains of the reactor buildings and wafting their contents into the atmosphere.[8]

CREDIBILITY MELTDOWN; RAYS OF RAGE

Officals' lying and even crying on television left the Japanese angry or despondent—not including the corporate cleanup-bidders. Apathy and the tradition of trickle-down trust drew few to public protests, although a few weeks after Fukushima several hundred met at Tokyo Electric's offices and marched to the offices of Japan's nuclear regulators, shouting, "Tokyo Electric, get out of nuclear energy!" "Compensate the victims!" Some demanded the company and government apologize. Others carried placards: "Even if we don't have nuclear power, we'll still have electricity." They suggested that the well-advertised insomnia of the first minister could be helped by waking up the premier. One participant, aged seventh, told the media who bothered to cover the events, "The Japanese usually don't protest, but now we must show that we can call for change." He had been protesting since March 12th, with only ten people at first, then three hundred at a later march. He promised to keep protesting until nuclear plants were shut down.[9]

Tens of thousands more were buffeted by fear and confusion. More than 80,000 threw together bare belongings and hit the road trying to run from the cloud of radiation that, due to wind shifts, followed them west and north. The things they carried included nearly all useful or edible. Those unable to flee had shambles, few improvised tools or nets but no money—no one would even consider visiting the area much less buying fish there. Some of the refugees' dire circumstances even put Tokyo Electric, government officials, and financiers at odds with their own bureaucracies. Tokyo Electric's first public dispatches that it would restart the reactors and build new better-sited, safer ones provoked outrage; some of its offices were even defaced by employees. "Tokyo Electric may want to ignore the feelings of Fukushima residents, but announcing plans to build new nuclear reactors is in no way acceptable," fumed Yoichi Nozaki, director general of Fukushima's Planning and Coordination Department, overseer of energy matters in the regional capital. "It is just unbelievable."[10]

Meanwhile, the eight-mile exclusion zone boundary proved porous; many were allowed to return to recover personal belongings, some made no secret of defiantly staying put. Kiyoshi Abe, a farmer in Minamisoma, about eight miles from the nuclear plant, spoke proudly about staying home and disparaged runners: "I'm amazed the Japanese are so obedient," he said.[11] As in Chernobyl, an increasing number of the less obedient slipped back into the zone to reclaim or steal what they might find buried or hidden. Others took up night fishing or farming rather than exile. They endured, as many do now, in a dark, radioactive no-man's-land of fear and lack of electricity, gas, safe water, navigable streets, or adequate shelter. A long radioactive rain began as well.

In a few weeks, the regime ordered some back into the zone to check that no one was farming on too large a scale and that herdsmen had executed all their stock. Many residents of Minamisoma, eighteen miles north of the Daicihi reactors, returned to retrieve some possessions and perhaps do some cleanup, but were shocked and shuffled around by Self-Defense soldiers and grim men with Geiger counter so much that they mainly wept or stumbled around. They saw what had been a tourist town of 70,000 with a popular annual samurai festival turned into wreckage dump, with crushed buildings, cars, and boats thrown around and a swamp of slop. The cowed group probably had their vision somewhat softened by the masks they donned—uselessly—to protect themselves. They hefted, hid, or buried what belonging they could find and made plans where to send their children; some of them met the nuclear hermits whom they encountered in the ghost town—not many since those who had stayed secreted themselves most of the day from radioactivity or meddling officials. Masks and cave life might have reduced some exposure, but invisible bullets had much else to target.[12]

DEBATE GOES GLOBAL AND LOCAL

Fear of widespread pollution prompted a weighty United Nations debate. Its upshot was that the International Atomic Energy Agency recommended that the world add to its 440 nuclear reactors, now generating about 14 percent of global electricity, and announcement of plans for or actual construction of new, supposedly safer, plants in India, China, and other countries around the world. Little mention was made of destroyed, decrepit, or middle-aged reactors, many of which were being fired up for more power production or were still in service.

The debate took place at the fifth Convention on Nuclear Safety April 4, in Vienna. As the debate spun on, the Japanese government continued dumping radioactive water and melt from Daiichi into the sea. It gave short shrift to Philip White of the Tokyo-based Citizens' Nuclear Information Center and Jan Beránek, director of Greenpeace International's Nuclear Campaign. Beránek argued the IAEA was preventing an honest review by "focusing its efforts to restore the public confidence in nuclear power and to help other countries expand the usage of civilian nuclear reactors to generate electricity."[13] The same could be said of national so-called nuclear regulatory agencies.

Some lip service followed. Cabinet Secretary Edano said Tokyo Electric was trying both old and new ways to halt the highly radiated fluids still flowing from a collecting pit cracked under Daiichi. He briefly tried on the robes of a prophet of doom; "If this situation continues for a long time, the issue of accumulated radiation contamination will increase. Even if the sea does disperse the radiation, it will become a bigger problem, so we need to stop this from spreading out to sea as quickly as possible."[14] But not to worry. To its defensive arsenal of sawdust, blotting paper, and shredded newspaper sops, Tokyo Electric was now trying to neuralize pollution with chemicals used in diapers.

Even before the great debate, Tokyo Electric had sought go-aheads from prefect and local officials for building two new nuclear reactors in Fukushima. The plan was included in a formal report on Japan's future power needs submitted to authorities in Tokyo on March 31. Construction on the "supplementary" reactors would begin in spring 2012.

Incredulous and enraged, Yoichi Nozaki, director general of Fukushima's Planning and Coordination Department, and Fukushima province's governor Yuhei Sato responded that "we definitely cannot accept" the building of new reactors. They delivered this opinion directly to Tokyo Electric's provincial branch manager and to its traveling "truth team." Tokyo Electric went over the heads of the locals and submitted unaltered plans to the Ministry of Economy, Trade and Industry. In the process the waves of anger from Fukushima brought forth a revealing admission from a power company official in Fukushima, who allowed that the reactor plans might well have been revised, since they took shape before the March 11th tsunami.[15]

INDIA CASTS THE FIRST STONE

Three weeks post-disaster, India's Ministry of Health and Family Welfare banned food imports from all Japan The ban applied until the Ministry was satisfied that the radiation hazard had subsided to acceptable limits. Other countries sanctioned dairy products, meat, fish, and produce from areas near the crippled nuclear power plant; many others, contemplating bans, began microscopic examinations of radiation levels in imports from Japan. Even Japan considered imposing radioactivity restrictions on seafood after contaminated fish were found in seas hundreds of miles south of the damaged and leaking reactors.[16] The U.S. has banned all foods that Japan previously prohibited for intra-national sale and the EPA policy announced in 2013 was to inspect radiation levels in any foodstuff from anywhere in Japan. However, the website Fukushima Diary points out that the U.S. safety standard is 1,200 Bq/Kg (Cs-134/137), twelve times higher than the Japanese limit.

The U.S. EPA was also moved to establish Maximum Contaminant Levels for concentration of radioactive iodine and other radionuclides in drinking water from public water suppliers as it issued environmental standards for the proposed permanent waste dump inside Yucca Mountain, Nevada, eighty miles northwest of Las Vegas. The water standard limits the radiation exposure of individuals, and radionuclide concentrations in groundwater from the release of I-129 and other radionuclides in the vicinity.[17] More controversial than Japanese imports or water issues is use of Yucca Mountain for creating what would be the world's largest nuclear waste facility. Protests have forced President Obama to delay construction. His opponents point out that keeping waste in temporary facilities will cost taxpayers (not nuclear plant operators) mega-millions.

At any rate, a growing number of people in the U.S. and elsewhere trust grocery chains and tap water—and government protection—about as much as a Las Vegas slot machine. Air? Worry when the flight attendant asks how you'd like your iodine pill.

NOTES

1. Committee to Assess Health Risks from Exposure to Low Levels of Ionizing Radiation, National Research Council, *Health Risks from Exposure to Low Levels of Ionizing Radiation: BEIR VII Phase* 2 (Washington, DC: National Academies Press, 2006), http://www.nap.edu/openbook.php?record_id=11340.
2. U.S. Environmental Protection Agency, "Cesium," *EPA/Radiation Protectio*n, March 6, 2012, http://www.epa.gov/radiation/radionuclides/cesium.html.
3. Ken Ijichi and Moshe Komata, "Efforts to Plug Japanese Reactor Leak Seem to Fail," *New York Times*, April 3, 20011.
4. Risa Maeda and Yoko Kubota, "Japan Releases Radioactive Seawater, Uses Bath Salts," *Reuters/U,S,*, April 4, 2011, http://www.reuters.com/article/2011/04/04/us-japan-fortimeline-aprilfour-idUSTRE7334C220110404.
5. U.S. Enviromental Protection Agency, "Cesium," *EPA/Radiation Protection*, March 6, 2012, http://www.epa.gov/radiation/radionuclides/cesium.html. Ibid.
6. Jacques Cousteau, "Periodic Essays: The Bounty of the Sea," *Deviant Art*, 1994, http://lemon-sauce.deviantart.com/journal/The-Bounty-of-the-Sea-by-Jacques-Cousteau-231816293.
7. Nuclear Regulatory Commission, *Reactor Safety Team, Confidential Assessment*, March 26, 2011.
8. James Glanz and William J. Broad, "U.S. Sees Array of New Threats at Japan's Nuclear Plant," *New York Times*, April 5, 2011.
9. Ken Ijichi and Moshe Komata, "Efforts to Plug Japanese Reactor Leak Seem to Fail," *New York Times*, April 3, 2011.
10. Andrew Pollack, "Japanese Revisit Nuclear Zone While They Can," *New York Times*, April 21, 2011.
11. Andrew Higgens, "Amid Nuclear Crisis, Japan's Tepco Planned New Reactors," *Washington Post*, April 6, 2011.
12. John M. Glionna, "Anger and Abandonment in a Japanese Nuclear Ghost Town," *Los Angeles Times*, March 31, 2011.
13. Jan Beránek, Yukiya Amano, and Phillip White, U.N. Nuclear Watchdog Says It Will Continue to Push for New Nuclear Power Plants Despite Growing Global Nuclear Concern, Democracy Now, April 5, 2011, http://www.democracynow.org/2011/4/5/un_nuclear_watchdog_says_it_will.
14. Yukio Edano, Japan to Release 11,000 Tons of Radioactive Water into Ocean, *Democracy Now*, April 4, 2011, http://www.democracynow.org/2011/4/4/headlines#441.
15. Andrew Higgins, "Amid Nuclear Crisis, Japan's TEPCO Planned New Reactors," *Washington Post*, April 5, 2011.
16. Ratnajyoti Dutta, "India Bans Japan Food Imports, Says Radiation Spreading," Reuters/New Delhi, April 5, 2011.
17. U.S. Environmental Protection Agency, "Iodine," *EPA/Iodin*e, March 6, 2012, http://www.epa.gov/radiation/radionuclides/iodine.html.

Chapter Nine

Nature as Nuclear Trash Bin

Tokyo Electric Power Company faces as much as 2 trillion yen ($23.6bn; £14.5bn) in compensation claims, according to JP Morgan. As the company struggles to contain the nuclear crisis, its shares have taken a beating on the Tokyo stock exchange, having lost more than 75 percent of their value since 11 March and reaching record lows. A key issue concerning damage compensation is whether the Fukushima nuclear plant accident is considered an unavoidable natural disaster. Investors have been worried about the impact of the current crisis on the company's profits, but the electric company may have an out. Under Japanese law, operators of a nuclear facility can be waived of liability if an accident is deemed to have been triggered by a natural disaster of an exceptional character. The Japanese government is yet to decide whether it will classify the earthquake and tsunami as an exceptional disaster, but arm-twisting is under way.[1]

In early April 2011 Japan Maritime Self-Defense Forces found high levels of radioactive isotopes, including cesium-137 and iodine-131—the iodine spiked to 3,355 times the legal limit for seawater, an indication that more radiation from unknown sources was entering the ocean. Cesium was also found to be twenty times its safety limit on March 28th.

Radiated seawater can injure ocean life in several ways—by outright kills, by creating "bizarre mutations" in their offspring; or by passing radioactive material up the food chain—here is a brief list of possibilities:

- Mortality and fertility reductions of marine animals off Fukushima shores. If radiation leaks continue, Japan could be dealing with marine genocides. The coastline isn't Chernobyl—it is impossible to isolate the sea or cement over the whole northeastern seacoast.

- Mutagenic changes or damage to reproductive systems of sea life. Marine organisms' eggs and larvae are highly sensitive to radiation, since radioactive atoms can replace others in their bodies and alter their DNA. Most deformed organisms die, but some live to pass on abnormalities. Either way, radiation would lower the ability to survive long-term.
- Massive die-off and mutagenicity of soft-bodied invertebrates such as jellyfish, sea anemones, and marine worms—which absorb radiation more quickly than shelled creatures (See "Chernobyl Birds' Defects Link Radiation, Not Stress, to Human Ailments.")
- Marine life burned and injured marine life after passing through pockets of radioactive material.
- Disappearance of predator species in the Pacific such as tuna and sailfish already stressed by overfishing—especially susceptible during spawning season.
- Persistent presence of radiated sea water. Though the half-life of iodine is only eight days and its isotopes present only a short-term problem, cesium takes thirty years to decay and plutonium thousands of years. Oceans can dilute radiation, but one cannot bury or halt the spread of sea radiation.
- Radiation moving up the food chain when animals or humans eat irradiated plants and animals. Seaweed and kelp, used in foods and fertilizers, quickly absorb iodine.

ANOTHER CHERNOBYL?

After Chernobyl, long-term evacuation and food restrictions were ordered in areas with radioactivity above 0.55 MBq m–2, but some believe this limit was too high and would be too brief.

On the basis of the Fukushima data collected in March 2011, it appeared that food restrictions there could remain for decades, particularly for wild foodstuffs such as mushrooms, berries, and freshwater fish.[2] That is without taking into account continuing leaks and the dump of wastewater into the ocean. The so-called liquidators of Chernobyl sent in to clean up farmland and decontaminate empty zones were cited for ineffectual effort and spotty success, prompting the 2014 efforts to put a steel lid on the reactor there.

In contrast, experts in the British isles looked for more extensive measures of prevention and repair. The UK Health Protection Agency's *Recovery Handbook for Radiation Incidents* details a range of verifiable cleanup measures for residential areas, including removal of topsoil and resurfacing of roads. British officials showed themselves highly interested in reactor design and operations—and safety protocols that had to be strictly maintained, especially since more reactors were scheduled to go online. They warned of increments in risks in step with successes in using nuclear energy. "There can be little doubt," the *Sixth Report of Royal Commission on Environmental Pollution* forecast in 1967, "that as the number of reactors in use throughout the world increases, there will in time be a major accident. Such an accident, even if it occurred abroad, would certainly cause anxiety in this and other countries and might require reactors to be shut down while safety aspects were assessed. This alone suggests that it would be prudent to aim for diversity in energy supplies."[3]

In Japan remediation sometimes took on aspects of voodoo. "Experts" with limited evidence and no experience advised spreading potassium on crops to compete with radio-cesium uptake, and giving "Prussian blue" boluses to grazing animals to reduce radio-cesium absorption. Such measures would be costly and might even add to the presence of contaminated waste. Consumers probably would refuse products from contaminated areas even when they meet regulations

Even today, without authoritative health data, the serious impact of Chernobyl has been noted, though officialdom tends to stress psychological rather than health consequences.

- One 2006 estimate by the International Agency for Research on Cancer foresaw 16,000 cases of thyroid cancer and 25,000 other cancers resulting from the radiation, among "several hundred million cancer cases from other causes." An individual's risk might be relatively low, but the extent of the disease would reach all.
- In 1991, however, a study by the International Atomic Energy Association set up as part of the "Atoms for Peace" effort to promote safe, secure and peaceful nuclear technologies, found psychological effects to be "wholly disproportionate to the biological significance of the radiation." The study undertook to provide accurate information about radiation health risks to affected populations in the thirty miles around Chernobyl.
- Fifteen years later in 2011, the UN Chernobyl Forum Report still concluded that Chernobyl's impact on mental health is "the largest public-health problem caused by the accident to date." It dwelt on misperceptions and inefficient compensation, that it said have led to widespread fatalism and feelings of victimization among locals. It concluded that rises in alcoholism and smoking may have done more damage than radiation exposure.

At any rate, at Chernobyl, vastly more has been spent on physical remediation than on public engagement, screening, or treatment. The failure to solve social and psychological problems relates not only to a lack of effort but also what one report calls "the intractability of the problem."[4]

ECHOES OF CHERNOBYL

Japan only reluctantly, in May, raised the severity rating of its nuclear crisis to the maximum seven, also assigned to Chernobyl by a ranking devised by the International Atomic Energy Agency. Level seven designates major releases of radiation with widespread, immediate, and long-term health and environmental effects. Opposing the rating, Japanese *government officials* implied that Fukushima was no worse than the rated-five Three Mile Island accident in 1979. They made much of the fact that the Chernobyl reactor exploded while in use, while Fukushima exploded after the reactors went offline.

Japan's nuclear safety commission was somewhat more forthright, putting the Fukushima plant's reactors at 10,000 terabecquerels (10,000 trillion) of radioactive iodine-131 *per hour* into the air for several hours after they were damaged. One becquerel indicates that a radioactive atom is splitting and releasing possibly harmful rays each second. The commission, however, insisted radioactive launch from Fukushima was about 10 percent of Chernobyl's. The nuclear safety commission claimed emissions had dropped in mid-April to less than one terabecquerel per hour, adding that it was continuing to monitor, with robots and computers, the amount of released radioactive materials. Conflicting reports of damage, death, and desperate need frightened and confused people, especially in smaller towns, but assistance and accurate information oozed out slowly usually in the form of estimates and gestures, or sometimes with bootstrapping pep talks.

Journalist Tetsuo Jimbo wrote about the contradictory claims, denials, and neglect in Rikuzen-Takata, less than a hundred miles north of Daiichi. People first thought they had escaped disaster, but then the tsunami swept in, submerging salients in the churning sea, creating a vast scrap heap on the large lowland, flooding fields, and smashing wrecked cars and building parts against anything upright, as if taking bites before the sucking, chewing tsunami undercurrent. As the surge rolled, people were ordered to supposedly safe shelters, soon flooded and blocked by the water and what it carried. Half of the town's 70,000 inhabitants were presumed dead after the day of the tsunami; the other improvised some sort of survival in the shambles. Jimbo reported chaos and resistance to clueless authorities: "The government is overwhelmed. They don't have enough personnel to devote to respond to the relief or the rescue effort." Measuring radiation exposure was low on the list anyway of those sent to survey the flattened, flooded town, soon receiving aftershocks and ocean and air after-poisoning from the seething mess to the south.[5]

After weeks went by and radium-dioxide explosions at three of the Daiichi reactors became probable, the government finally extended the evacuation zone to include five settlements outside the twelve-mile no-go area.

Yukio Edano said that at twelve to fifteen miles, people would be safe and the radiation inside the zone would dissipate or be cleaned up. Thomas Breuer, head of the Climate and Energy Unit for Greenpeace Germany and part of a field team of radiation monitors, strongly disagreed.[6]

He pointed out that Fukushima was truly no Chernobyl in one aspect: the Russian plant meltdown occurred in a sparsely settled area on a relatively flat plain. The Fukushima bombs went off by the sea and in a densely populated area. He called for more and further evacuation: "We warned the government that there are a lot of cities and villages outside the 20-kilometers evacuation zone where the radiation levels are so high that people need urgently to be evacuated, especially children and pregnant women, because they are the most vulnerable part of the population to radiation."[7]

Paul Gunter of Beyond Nuclear predicted that worse developments were on the way. Thousands had been radiated and refugees were packing reactive and reacting items across country. The nuclear plants were ready to fall to pieces, shaken, cracked, undermined, burned. Some of the protection efforts were inept, if not ludicrous, such as advising people to use diaper chemicals for protection, or the Keystone Kops of the air, the water bombers who missed more targets than they hit and, anyway, added to pollutants running into the soil and sea. Said Gunter, "The walls of defense are falling, with the melting of the cores, the collapsing . . . we're expecting the collapsing of the vessels. And then, with these damaged containments, these are all open windows to the atmosphere."

It can't happen here? As in Japan, the U.S. atomic oversight is generally in the hands of pro-nuke partisans who claim U.S. reactors and waste-tending are safer than Japanese. As in Japan, the media, source of most public opinion, leaves such claims unquestioned, atomic issues unexplored, instead beating the voodoo drums of consumerism and technological Disneylands. But Karl Grossman noted that even the Nuclear Regulatory Commission itself in 1985 put at 50 percent the chance of a severe fuel core accident in one of more than one hundred nuclear power plants in the U.S.[8] So far, regulation of the nuclear industry is mainly in the form of requests for self-regulation.

THE SHELL GAME OF DELUSIONS

At the end of May 2011, the U.S. Department of Energy aerial survey of radiation doses at Daiichi lent more weight to the extent of damage—and to the usual official sidestepping.

The DOE found a zone 25 miles northwest of the Daiichi plant with a dose rate above 125 microsieverts per hour, a level at which instant evacuation is recommended. (One gets 20 microsieverts in a chest X-ray, 240 microsieverts annually from food). Some types of radiation were declining as isotopes with short half-lifes decayed. But, as with the 1986 Chernobyl accident, cesium-137, with a half-life of 30.2 years, will determine the long-term impact on the contaminated region and its residents.

In May also, the International Atomic Energy Agency (IAEA) reported severe danger to human life—cesium-137 deposition ranging from 0.02 to 3.7 megabecquerels (1 million splitting and radiating atoms per second) per square meter thirty miles from Daiichi. At Iitate village, twenty-four miles northwest of the plant, Japanese teams found radiation at higher levels. Perhaps surprisingly, there is in 2014, still no clear information on cesium-137 contamination within the evacuation zone, although the DOE estimates it could be of the order of four megabecquerels a square meter, if the isotopic composition of deposits near the plant is similar to that in the area farther to the northwest.

Said the IAEA, "The implications of these data are far-reaching and socially wrenching. If large areas are contaminated with 0.5 MBq m–2 or more, evacuation could be for the long term." The slippery language—"long term"—seems more inaccurate than pragmatic.

The Japanese authorities would have to rewrite the rule-book on clear-and-present emission dangers, as they had begun to do in allowing doses of 250 mSv for radiation workers and even children in the twelve-mile zone. After an accident, it may be appropriate to set exposure limits for members of the public higher than the typical 1 mSv per year maximum. But a limit of 5–10 mSv per year (perhaps with voluntary resettlement at doses above 1 mSv per year) may be appropriate, for the owners of Daiichi, bearing in mind that millions of people in areas of high natural radioactivity worldwide already are exposed to more than 10 mSv per year, and that occupational exposures (for example, to long-haul air crews) can be around 5 mSv per year.[9]

NOTES

1. "TEPCO may face $23.6 billion in claims, JP Morgan says," BBC News, April 12, 2011.
2. J. T. Smith et al., "Pollution: Chernobyl's Legacy in Food and Water," *Nature* 405, no. 6783, Brief Communications (May 11, 2000), http://www.nature.com/nature/journal/v405/n6783/full/405141a0.html.
3. *Sixth Report of Royal Commission on Environmental Pollution: Nuclear Power and the Environment* (London, England: Her Majesty's Stationery Office, 1976).
4. Mark Peplow, "Chernobyl's Legacy: Twenty-five Years After the Nuclear Disaster, the Clean-up Grinds On and Health Studies Are Faltering. Are There Lessons for Japan?," *Nature*, March 28, 2011.
5. Juan Gonzalez, "Report from Sendai: Fears of Radioactivity Are Hampering Relief Efforts," *Democracy Now*, March 17, 2011, http://www.democracynow.org/2011/3/17/report_from_sendai_fears_of_radioactivity.

6. Justin McCurry, Tokyo, "Nuclear and Industrial Safety Agency Raises Severity Rating to Maximum on International Nuclear and Radiological Event Scale," *Guardian*/UK, April 12, 2011.

7. Amy Goodman, "Nuclear Catastrophe in Japan "Not Equal to Chernobyl, But Way Worse," *Democracy Now*, April 12, 2011, http://www.democracynow.org/2011/4/12/nuclear_catastrophe_in_japan_not_equal.

8. Ibid.

9. Jim Smith and Nicholas A. Beresford, *Chernobyl: Catastrophe and Consequences* (Berlin, Germany: Springer/Verlag, 2005).

Chapter Ten

Can Anyone Ever Pay?

Nature and species life are not for sale. Some projections of the NRC put the value of a human life at $1 million, but no monetary compensation, however large, can ever be enough to offset, rectify or rebuild after ecocide or genocide. Yet after the nuclear chaos in Japan, some seek the courts as the only means of redress. Will Tokyo Electric and the Japanese government be put in the dock and charged with mass depredation and what amounts to murder? If so, will the government use the Japanese population's resources to bail out or nationalize TEPCO, so it becomes the primarily responsible party to pay compensation for destruction of essential natural conditions and the irradiated Japanese and others? Will liability be extended to the owners of the Rokkasho plutonium fast-breeder plant, General Electric and Areva, which produced the plutonium-uranium fuel MOX for the Fukushima Daiichi Nuclear Plant and others around the world?

Litigation will almost certainly arise from both governments and firms in Russia, Korea, China, and other countries, whose oceanic resources have been compromised and will continue to be contaminated by the Daiichi nuclear disaster for many years to come.

In the immediate future, there is no end to rumors about the danger of Japan's marine and farm life. Is genetic damage to the sea's feeding chain soon to reach the human species? Will farm output lose sales, regardless of the extent of the content of radioactivity in their products?

These are among the many hundreds of questions raised by the explosions and nuclear bullets released at the Daiichi nuclear plants.

HOW HIGH THE PRICE?

Tokyo Electric Power Company may face as much as 2 trillion yen ($23.6bn; £14.5bn) in compensation claims, according to JP Morgan.

Under the plan, a new fund would be created with mandatory contributions from electric utilities, including Tokyo Electric, in case total compensation awards exceed its financial capacity. The government could add public money if needed. The utility would be required to repay any money it uses from the fund; in May it agreed to drastic restructuring, cost-cutting, and other conditions.[1]

As the company struggled to contain the nuclear crisis, its shares took a beating on the Tokyo stock exchange, losing more than 75 percent of their value between 11 March and May 5th, in the process hitting all-time lows.[2]

TOKYO ELECTRIC SIDESTEPS

Given the probability that Tokyo Electric would be unable to compensate Daiichi victims, the nuclear-lobbied Japanese government approved a May 2011 multibillion-dollar bailout, whose costs would be borne by taxpayers being served with oil-generated and thermal power. The industry would be largely left to self-regulation. Taxpayers would pay not only higher taxes but higher utility charges for the power generated more cheaply and safely than by the use of nuclear fuel. Tokyo Electric would have access to a quasi-bank from which funds could be drawn as compensations were (slowly) approved—the funds to be returned to the government at some later unspecified date. The government would issue bonds to raise funds for the bailout and, as well, require all nuclear power companies to contribute to a bankruptcy insurance account. The government's setting itself up as a hedge fund began to cool reaction to and doubts about the corporate-government complex.

In return for government help, Tokyo Electric, accounting for some 33 percent of total electricity of Japan's economy, agreed to massive cost-cutting, no upper limit for compensation payouts to victims, and making managers subject to investigations that might humiliate them but not put them behind bars or into bankruptcy as it turned out. They produced symbolic tears, kowtowing in Tokyo Electric uniforms, well-acted sincere apologies, and faced nominal penalties. The company chair resigned with regrets and a severance package.

Despite infrastructure damage, leery consumers, and rising unemployment, the demand for electricity is healthy enough that Tokyo Electric may raise needed funds from higher electricity prices. In effect, the utility would impose a super tax on the nation's consumers of power, including transportation, medical, and other institutions, industry and agriculture, and, especially, on individual electricity users. Investors and creditors were dubious and displeased, as evidenced by a drop of about 6 percent for TEPCO shares and collateral reductions of stock prices for banks backing or holding utility shares.

To bridge the supply-demand gap, the power company returned to service old warhorses fed on oil and coal. As the hot summer of 2011 came on, Tokyo Electric fired up its Hirono plant about twelve miles south of the Fukushima plant, revving up its oil-powered 600-megawatt No. 2 unit and 1,000 megawatt No 4 unit. It readied for operation a 1,000-megawatt coal-fired unit at Hitachinaka in Ibaraki, north of Tokyo.

Against the possibility of more accidents, nuclear power operators began shutting down aging and at-risk reactors, such as the Hamaoka nuclear complex 120 miles south-west of Tokyo and near a tectonic fault line, regarded by seismologists as a geologic time bomb. Japanese ecology and economy might remain at risk, but Tokyo Electric, its backers and shareholders, anticipated smoother sailing.[3]

KEY LEGAL ISSUE: UNAVOIDABLE NATURAL DISASTER

Tokyo Electric may even avoid payouts for its missteps. At issue concerning damage compensation is whether the Fukushima nuclear plant accident is considered an "unavoidable natural disaster." Under Japanese law, operators of a nuclear facility can be waived of liability if the accident is deemed to have been triggered by a natural disaster of an exceptional character. The Japanese government is yet to decide whether it will classify the earthquake and tsunami as an exceptional.

Now the situation in Japan has seen it again rise to the top of many political agendas. Since nuclear power produces no greenhouse gases—but does introduce large amounts of heat warming Earth—proponents are touting it as a green alternative to fossil fuels. They argue that one pound of enriched uranium provides the same energy as 3 million pounds of coal or 1 million gallons of gasoline. They neglect to say that nuclear reactor waste or emissions after an accident can cause cancer, birth defects, and mutant spawn. The technology is fascinating but ominous.

Nuclear power plants are complexes that span square miles. On a subatomic level, the sole purpose of a plant is to harness the energy of nuclear fission—a reaction whereby an atom's nucleus splits into two smaller nuclei, inducing nuclear fission in enriched uranium oxide, comprising 96–97 percent uranium-238 and 3–4 percent uranium-235. Uranium, the heaviest of all natural elements, is one of the easiest to break apart. When a relatively slow-moving free neutron runs into a uranium-235 atom, the atom will absorb the neutron, and the extra energy will make the atom unstable. The atom immediately splits apart, into two smaller atoms and 2–3 free neutrons. A fraction of the atom's original mass thereby becomes energy, in the form of heat and high-energy photons called gamma rays.

The right mix, or uranium-235, produces a chain reaction. Some of the free neutrons generated in the fission reaction encounter other uranium-235 atoms, causing those atoms to split apart, producing more free neutrons. Collectively, the splitting atoms generate substantial heat. All the equipment in a nuclear plant has one core function: harnessing heat to generate electricity.

To prevent the heat from rising, when water stopped circulating at Fukushima's Boiling Water Reactor (BWR) station, a buildup of hydrogen gas blew off the roof of the building. The fear of a radiation leak occurred when the coolant water (which immerses the fuel rods) failed and exposed the fuel in the reactor vessel to air. Control rods were supposed to slide in between the fuel elements to halt the nuclear process, but when power failed and the fuel elements flamed, radiation leaks began and technicians tried seawater to cool the fuel rods.[4]

SCOUTING OUT LAWYERS AT THE BAR

Shaken politically if not empathetically by new quakes, aftershocks, and reactor safety problems, the Japanese government put on a show of concern: it issued to nuclear power companies regulations that most knew were difficult to monitor and impossible to implement. It ordered plant operators to store at least two emergency backup generators for each reactor and install fire pumps and power supply vehicles as further precautions.

Objections and foot-dragging followed April 7, with more nuclear crises roiling the populace and the economy, the government ordered thirteen nuclear plant operators to check and improve outside power links to avoid earthquake-related outages that could cause safety systems to fail. More power outages during a strong 6.9-magnitude earthquake and aftershock on April 7 gave safety orders more punch if not compliance.

The demands were impossible anyway. The government was demanding plants reinforce the quake resistance of power lines connected to each reactor or rebuild them. They were also ordered to store all electrical equipment in watertight structures which would require moving the reactor hearts from their original positions below reactor floors. Thus had the GE design stipulated, wastes atop and electricity and motors below the reactor, exactly opposite what might have worked in the tsunami.[5]

All well-planned, no one at the upper reaches of the government regulatory agencies or at Japanese Electric had any intention of obeying the letter of the law or so-called regulations. Simply, they could not without putting the nuclear industry, perhaps the entire nation, into bankruptcy.

THE BEST-LAID PLANS

A blueprint for stopping radiation leaks and stabilizing damaged reactors within the next six to nine months seems an overly optimistic, if not dangerous first step toward allowing some of the tens of thousands of evacuees to return to the area.

The unveiling of the plan came two days after Tokyo Electric announced plans to give 1 million yen ($12,000) in initial compensation to each evacuated household, with much more expected later. On orders from Prime Minister Naoto Kan, Tokyo Electric drew up the blueprint and publicly explained its long-term strategy—for the first time since the disaster. "We sincerely apologize for causing troubles," Chairman Tsunehisa Katsumata said. "We are doing our utmost to prevent the crisis from further worsening." TEPCO planned to deal with the crisis in at least five stages:

1. Focus on cooling the reactors and spent fuel pools and reducing the level of leaking radiation. Decontaminate water that has become radioactive; reduce the amount of radiation released into the atmosphere and soil; and lower radiation levels in the evacuation area.
2. Control the release of radioactive materials; achieve a cold shutdown of the reactors; and temporarily cover the reactor buildings.
3. Establish a system to recycle cooling water to remove radioactivity as well as corrosive salt from seawater earlier used as an emergency cooling measure.
4. Remove fuel from the spent fuel pools and permanent, burial covers over the buildings.
5. Decontaminate the area, including removing and replacing the soil, so evacuees could return home.

Government officials fanned out across the affected areas over the weekend seeking to explain evacuation decisions and calm nerves, saying the evacuees would not have to stay in gymnasiums for such a long period, but would be moved into temporary housing. Government spokesman Edano met with the governor of Fukushima, who has vigorously protested the predicament the nuclear crisis poses for his prefecture. "The safety of residents is our foremost priority," Edano said. "I told the governor that the government will do everything it can to prevent the crisis from worsening."[6] So, as usual, he advised and offered little more than calm.

FLAWED OINTMENT

With an entourage of nuclear lobbyists, national officials met with local authorities to demark and harden twelve-mile radius "caution zones" around the power plant, now a matter of law, not a voluntary evacuation. More or less. In the usual absence of credible information, many thought radiation releases from Daiichi were finished and made brief incursions to their old homes to get items of value or need. Journalists, too, began poking through the zones, talking to returnees and dozens of families who had sat out the radio-run. One can imagine the hard-hitting probes, akin to those previously unable to pierce the tissue of lies and disinformation of corporate or government oil-spreaders: *Tell our viewers, in this time of national disaster—how do you feel?* Many evacuees feared that their vacant homes might be robbed. And regardless of irradiation dangers to the public, nuclear power experts seriously debated whether the continued movement of people, livestock, and vehicles would make it harder to decontaminate after the nuclear power plant had been fully stabilized. The people should go or stay—not be constantly returning.

Elderly, sick, and hard-bitten people outright refused to live at evacuation centers or school gymnasiums, some two hundred households stubbornly remaining in the reach of radiation. Many farmers left so quickly they forgot to feed or water their livestock, the bloody aftermath unbearable for those who returned to see their dead cows and goats tethered to posts or starved in their pens. Other farmers who refused to abandon their cows said they would rather share their fate and die with them. Acts of grief and bravery appeared everywhere.

Government plans for emission demarcation areas were tried, ignored, and finally accepted to allow residents to make brief trips into the zone to recover essentials from their homes, with strict warnings that a legally enforceable caution zone would otherwise be created. Such warnings were strictly illegal, unenforceable, and it turns out that even Fukushima Prefecture police, with jurisdiction over the evacuation zone, did not know who crossed the invisible barrier or even how many occupied houses were still within its confines.

Counting heads made the Japanese cabinet seem self-important, though none of these themselves ventured to enter the radiation theater. There had been 78,200 residents inside the twelve-mile radius exclusion zone before the earthquake, tsunami, and nuclear accidents. How many remained, no one knew precisely. Ignorance seemed to prevail among the census takers surveying another 62,400 people who lived between twelve and nineteen miles from the power plant. Authorities initially told people living inside the larger area to stay indoors, but thereafter asked them to leave voluntarily, along with residents of five other communities outside the zone that also received some radioactive fallout because of wind and rain patterns.

A problem with computer models of radio-pollution is their assumption of stable wind patterns and neglect of certain catchment areas or "hot spots' that appear after a nuclear power plant accident or leakage. These can be mapped based on wind velocities and direction, as well as rains or other precipitation concentrating radioactive fallout in culverts or catch-basins, near homes or public facilities, but computer reports alone are inadequate pollution projectors. On-the-ground inspection, a slow and sometimes inexact process, may not indicate the need to bury contaminated wood, cloth, and objects in landfills lined with sand and boric acid or encased in other protective material. Even stone, loams, and loose dirt may have to be dug up and promptly buried in permanently-sealed nuclear cemeteries.

Experts disagree whether activity and traffic in contaminated zones create new hot spots.

Michael Corradini, chair of engineering physics at the University of Wisconsin, reasoned that power crews setting up electricity transmission lines across the evacuation zone to the damaged power plant, along with heavy repair equipment being brought into the power plant, and the movement of private individuals and their vehicles would probably not have much additional effect in spreading out the hot spots.[7] The government, at any rate, has announced tighter control of evacuation zones, but failed to provide specifics. Recommendations, requests, and industry self-regulation remain the order of the day.

NOTES

1. Mari Yamaguchi, Tokyo, "Japan Nuclear Reactor Damage Worse Than Expected," *Huffington Post/HuffPost World*, May 12, 2011, http://www.huffingtonpost.com/2011/05/12/japan-nuclear-reactor-damage_n_860999.html.
2. "Tepco May Face $23.6bn in Claims, JP Morgan Says," BBC/Business/News, April 12, 2011, http://www.bbc.co.uk/news/business-13045327.
3. Roland Buerk, "Japan's Government Approves Tepco Compensation Scheme," BBC, Tokyo, May 13, 2011.
4. Jonny O'Callaghan, "Is Nuclear Power Safe?," *How It Works*, March 15, 2011, http://www.howitworksdaily.com/science/is-nuclear-power-safe/.
5. Mari Yamaguchi, "Radioactivity Rises in Sea Off Japan Nuclear Plant," Associated Press, April 16, 2011.
6. Ravi Nessman and Yuri Kageyama, "Japan Nuclear Crisis: Fukushima Plant Operator Announces Plan To End Crisis," Associated Press, April 17, 2011.
7. Keith Bradsher, "Japan Considers Tightening Access Around Nuclear Plant," *New York Times*, April 20, 2011.

Chapter Eleven

Returning—on Shaken Feet

About two months after the destruction of Daiichi, Tokyo Electric announced its plan for control and repair. No one believed it: in three months building new cooling systems to prevent catastrophic radioactive leaks. The company said it would temporarily cover three shattered reactor buildings and put in filters that cut back air contamination. It would cover three of the six reactor buildings at the plant and reduce venting. The temporary covers would be made of material like construction building-wrap fabric, though Tokyo Electric admitted that such covers could be torn in one of the common summer typhoons usual to the region.

During the cooling system building and tenting the plant, engineers would design more permanent containments of concrete. During the months of designing and applying protective covers, radioactive emissions would continue into the air and possibly the seas. Among other problems were the facts that the plant, the covers, could not endure using water heated by contact with hot metals and piping. Boiling and evaporation had to be avoided because fuel rods had to be kept underwater to avoid meltdown. In other words, some cooling efforts made things worse.

Especially the use of water. Tokyo Electric helicopters bombed the reactors and the pools that held still-hot used fuel rods with tons of water and sprayed them from giraffe-necked fire trucks, whose heads could reach over the jagged walls. But the spray turned to steam in the hot reactors, pressure built, so venting and steam carried more radiation into the air and condensing on land and sea. Hosing and venting, or "feed and bleed," added to the deadly threats, not only of poisoning but of blasts and burns.

The plant gushed radioactive materials into the air, the most dangerous, plutonium, potentially causing lung cancer to anyone who took even a breath of its vapor. Tons of runoff poisoned soils and sea—and every link in the food chain. Clouds and winds carried hot spots in the sky which came to ground in patterns unpredictable by the computer models and the guessing of Tokyo Electric, which would finally claim it was not responsible for where the poison landed—reminding one of the Tom Lehrer satire on Werner von Braun's claim that it was not his department where missiles came down; he only helped launch them.

In the face of gathering concern and public relations and mechanical pressures, Tokyo Electric announced a long-term (translation: slow to implement) solution. All-new cooling systems would eliminate runoff. They would form a closed loop—quite similar to the previous, failed, systems—fighting fire with fire, or hot water with cooler water. The systems would condense steam and pump the resulting fluid back through the reactor system rather than allowing escapes of moisture through vents or leakage due to pressure. During the months the system was built and installed, Tokyo Electric would filter radioactive particles and salt from reactor system water and store them or suspect gloop near the plant, in a water-dump structure of 30,000 tons capacity. It was assumed—no one even asked—that the waste walls would be as sturdy as those of Jericho.

Some risks won lip service. Goshi Hosono, adviser to Prime Minister Kan, asked if new cooling systems could be built quickly; and what might be the effect on the building or the site of an aftershock or turbulent tides. Hironobu Unesaki, professor at the Research Reactor Institute of Kyoto University, approved of the long-term plan but added that the company should try to achieve cold shutdown sooner than the planned six to nine months, to reduce risk of a massive radiation release. A long shutdown process would mean continuing danger for a large population who should have been evacuated instantly farther from the epicenter of the explosion, given the shaky schedule for shutdown and entombment of Daiichi.[1]

Radiation levels far too high for human exposure would slow progress, too. Spent fuel, once kept cooled in water on the top of four reactors, suffering hydrogen explosions, once roofs and walls had been blown away, sent fissionable particles into the air as far as a mile away. Only robots could find out how bad things were; they reported from Unit 1's reactor building that inside radioactivity readings of about 50 millisieverts per hour indicated a lethal atmosphere.

HAZARDING THE NUCLEAR MINEFIELD: SEND IN THE ROBOTS

In early May, workers ordered by Tokyo Electric into the forest of steel stalactites of Daiichi were like canaries in a nuclear mine. They were decked out in strange-looking protective gear, with masks attached to scuba diver air tanks. If these were heated to ignition point, the explorers would have become torches. They had a pledge of safety from Japanese authorities backing Tokyo Electric's experimental invasion and scarcely tested equipment. This was established by the official doubling of the legal limit of radiation exposure from 125 to 250 millisieverts a year, topped in a day in the littered pit.

The level exceeded the Law of Radiation Exposure set in the nineteenth century by Madam Curie, as well as the twenty-first-century standard of the U.S. nuclear industry of an upper limit of 50 millisieverts per year for nuclear plant workers. Exposed to the estimated 1,000-millisieverts-a-year levels in the plants, workers, just as the population at Hiroshima and Nagasaki, were set to die in three days of radiation that devastated their bloodstream and provoked diarrhea, nausea, and vomiting. Nearby trapped populations took some hope from Tokyo Electric's tent, the first visible sign that it was doing something. The tent kept off rain but did not prevent gamma rays passing its walls and showering on 80,000 relocated people in the twelve-mile (twenty-kilometer) radius, jammed into schools, gyms, and community centers.[2]

Radiation hobbled repair work or even registering danger at Daiichi. The exploratory robots themselves expressed shock, instrumental if not emotional, as they went off the scale trying to record levels of lingering radioactivity into which even Tokyo Electric hesitated to send regular work crews. Robots nudged reactor shards, opened doors, circled debris and puddles, and returned with temperature, pressure and radioactivity readings showing no trend toward even official safety levels. At Unit 1, robots detected up to forty-nine millisieverts per hour; at Unit 3, they read fifty-seven millisieverts per hour. Far higher readings had already come from areas where contaminated water pooled, as in the Unit 2 turbine building, suspected of cracking and leaking nuclear material.[3]

The robots were meant as pathfinders into Unit 1 for human workmen to connect ventilator devices to "absorb" radiation from the building air as a preliminary to installing a cooling system. Once this was done, Tokyo Electric hoped to create an atmosphere in which workers could install permanent cooling equipment to allow a safe, so-called cold shutdown. Radiation would have to be lowered to 5 percent of pre-ventilation levels—a point where management said it would be "comfortable" with opening open the double-entry doors and let in human surveyors and repairmen. About opening these doors to the outside, management said, "We expect only a small amount of radiation leakage in the immediate surrounding area and it shouldn't pose a threat." Tokyo Electric, in other words, decided what would or would not be a threat to the nearby ecology and population, just as would other utility companies operating nuclear behemoths.

The intentional venting created by opening the doors would clear the air, perhaps, in Unit 1, but more certainly expose miles around the reactor and the repair workers to radiation. The risks, though, were secondary, thought those in charge to the needs of the reactor complex: Eight hours after the doors were opened, workers entered the building to test radiation levels and began replacing the wrecked cooling system. Their lives, their way of life, their very history, thousands of deaths and diseases, trillions of dollars, and consequences yet unknown were parts of the price to be paid for what boils down, simply, to the creation of steam.

That is the main purpose of nuclear power plants, to boil water to create steam channeled into turbines which spin to generate electricity. To make the wheels spin, the fuel for heat, aside from flowing water, sun, geothermal or wind energy, has created both great profits and great problems. Daiichi's burning fuel rods to boil water collaterally seared the economic and financial foundation of millions of people. Not to mention medical or genetic issues, in ashes are their culture, regional identity, and connections to ancestral lands and memories. Many have taken the steps from distrust to disbelief and even on to dissent, but more seem prepared for sorrow rather than taking to the street.

SADDER. WISER?

Tokyo's Electric's nuclear plant seemed, at first to add to the quaintness and livability of Fukushima prefecture. It created a strange mix of tiny villages, farm communities, grazing cows and goats, and industrial construction and activity. In an area lacking natural resources, power plants could be promoted as good neighbors as well as contributors to employment and national security, as well as essential to secure future energy sources and commerce in metropolitan areas. Giant utilities would cure the ills of high unemployment, a shaky economy, and even improve education and skills of young people, whose paucity of such rivaled Okinawa, long the worst in Japan.

UNKNOWN FUTURES

Tectonics, the tsunami, and the nuclear cloud, however, darkened the bright hopes and left mere survival as the people's primary purpose. Human error and arrogance, greed, revealed to the suffering victims, may strengthen the determination of those ready to overturn subservience to Western energy colonialism, but public opinion, like the undersea quakes and the effect of the disaster, so far remains something of a mystery. What is known is that rebuilding and clean-up efforts, funded by government levies and higher taxes, as well as by higher electricity rates for families, go forward. The record makes clear that radiation and ruins remain, that the victims of the disaster are largely on their own and must face added burdens of expense, exposure, and isolation, at the same time rained upon by lies, mis- and disinformation, distractions of nationalism and wounded national pride, and the perpetual feel-good messages of national media and paid-off propagandists.

The following account details some of these short-sighted pseudo-solutions, the move to bury nuclear dangers past, and the turn to reconstruction.

NOTES

1. Ken Belson and Steve Lee Myers, "Tokyo Utility Lays Out Plan for Its Reactors," *New York Times*, April 17, 2011, sec. Asia/Pacific, http://www.nytimes.com/2011/04/18/world/asia/18japan.html?pagewanted=all&_r=0.
2. Tomoko A. Hosaka, "Workers Enter Damaged Reactor Building for 1st Time Since Right After Earthquake," *Huffington Post/HuffPost World*, May 5, 2011, http://www.huffingtonpost.com/2011/05/05/japan-nuclear-crisis-workers-enter-reactor-building_n_857880.html#.
3. Hiroko Tabuchi, "Radiation Poses Barrier to Repair Work at Plant," *New York Times*, April 18, 2011.

Chapter Twelve

Popular Rights vs. Corporate Power

In any nation, moments come when its people reject government spokesmen and wealth-driven vested interests and their in-house experts spinning fictions.

Such a moment is on the way in Japan, where owners of nuclear power plants place financial goals above the safety of the population and refuse to follow the changing political demands of politicians fearful of being turned out of office if they do not respond to outraged citizens and communities.

SUDDEN REVERSALS

The public's coming of age in Japan has much to do with its "economic miracle" of the last fifty years. Japan Inc. drew upon the general technical skills and industriousness of its labor force—and upon nuclear reactors for one-third of its increasing electricity requirements. Much of the technical miracle was merely finding ways to make water hotter, faster to turn steam-driven dynamos, one of them being the use of recycled plutonium waste.

The fuels, such as plutonium, used to boil the water to steam were hotter than steam itself and, moreover, unsteady amalgams of atoms that would absorb or lose electrons and emit rays in the process. So power companies had to pay attention to containment of the rays and keeping things cool.

Doing things as cheaply as possible, Tokyo Electric used forty-year old design plans for the nuclear structures and sited them close to the ocean, and, incidentally near a geologic fault line where land was cheap or could be gained by measures resembling eminent domain. The U.S. Atoms for Peace program, French suppliers of a more powerful fuel mix of plutonium and uranium, and the Japanese media, nuclear agencies, and government assured the relatively sparse population of the northeast that power production would be safe and economical, although in Japan the "too cheap to meter" phrase did not attain its level of usage in the West.

The Japanese had heard of Three Mile Island and Chernobyl, but these did not shake their faith in themselves as unique, smarter, harder-working, and with better leaders at the top of hierarchies of business and government than other nations. Japanese technology looked to become dominant in global markets; even its main competitor was snapping up products from Sony, Hitachi, Toyota and other economic stalwarts. Then came the quake and walls of water.

Among other aftershocks, the denials, damage, and fumbling cracked cozy certainties as awareness of the cost and continuing danger spread. At the same time the nuclear plants were releasing waves of invisible bullets passing through and destroying flesh and bone, taking away the very breath of Japan's people—creating terrors of inhaling nuclides that could destroy lungs and thyroid glands, shrivel the development of infants and children, and bring on an unfolding sweep of cancers and death. Government and power company measures often made things worse, some of them almost laughable, such as advising people whose homes were in shambles to stay indoors or trying to use paper blotters to clean up the sea.

Many protested against the inaction and ineptness; more tuned out prattling politicians covering up nuclear company lies parroted by the tightly held media. Politicians found that nuclear power companies ignored the rising demands of the public for assurances of safety, and they feared being caught between public outcries for truth and nuclear power company drives for financial returns. These, major contributors to gaming elections, thought to hush the uproar with their old bogeyman: absent nuclear power it would be lights out for Japan and a nosedive for the Nikkei average of widely held stock shares. The Nikkei did decline but not the anger of the exposed and exasperated.

LATE ADMISSIONS OF EMISSIONS

Goshi Hosono, Prime Minister Kan's adviser handling the Fukushima crisis, announced in late April that innovative methods of water decontamination were under study. The idea was to recirculate the fuel-cooling water instead of flooding the facility. Tokyo Electric also finally admitted the nuclide emissions from the explosion and partial meltdown of the Daiichi plant were sweeping outwards. It tried the cooling experiment of using water leaked from the nuclear core to fill and cool the containment vessel. It assumed the container was intact, that no chunks of melted fuel had fallen and damaged the vessel bottom.

But on May 3, entering the building for the first time since March 12, repair workers' gauges showed the water level was at least three feet below the thirteen-foot-long fuel rods suspended in the vessel. Some rods had melted, as had happened at Chernobyl; but power officials assumed chunks and melt of damaged fuel rods were sitting at the bottom of the vessel, covered in water.

But what if they were not?

The low level of water already indicated that the core of Unit 1 had a bigger breach than expected. Cooling water also had been leaking from the reactor cores of Units 2 and 3, allowing an estimated 70,000 tons of contaminated water to pool inside the complex. Only on May 5th, did Japanese authorities reveal that one of the reactors at the crippled Fukushima Daiichi nuclear power facility suffered a partial meltdown of its core—the nuclear disaster was worse than previously disclosed.

Repairs to monitoring equipment revealed the new data, which also showed that the water level in the core of Unit 1 at the Fukushima Daiichi plant was much lower than previously thought, leaving parts of the fuel rods fully exposed. Other fuel and melt slumped to the bottom of the pressure vessel which Tokyo Electric said were most likely covered in water.

After-0the-fact disclosures were both shocking and reminders that the government had lied from the start. On May 17th, the utility released data showing some of the fuel rods had *completely melted.*

Reactor Number One had automatically halted operations immediately after the 2:46 p.m. quake. But the water level in the reactor dropped and the temperature began rising at around 6 p.m. The damage to the fuel rods had started by 7:30 p.m., with most of them having melted by 6:50 a.m. the following day. The building then exploded at about 3:36 p.m.

Goshi Hosono disclosed more than two months later on May 16th that reactors 2 and 3 may have experienced similar damage. He was almost reassuring with his report that power company data and analysis revealed the fuel rods went uncooled for more than six hours. It was a learning experience: "We must learn from our failure to recognize that Reactor One's core had melted."[1]

The learning experience had radiated many thousands of nearby resident communities. Using evacuation routes mandated by the government, many would receive additional dangerous doses.

WITHOUT WARNING

Greater-than-expected leaks in Daiichi plant vessels also had sent radioactive water pouring from troubled reactors into pools around the complex, obstructing repair and control efforts. No warning went out to the population within the nuclear fissionable range, although an energy official admitted radioactive water had leaked from the reactor and the exposed fuel rods had melted. Surges of radiation were hitting mop-up crews and people within thirty to fifty miles of the seething complex.

Why is this a danger?

No one knew what to expect. Then recent temperatures inside Unit 1's core were at the most 237 degrees Fahrenheit (114 Celsius), well below the normal operating temperature of about 570 Fahrenheit (300 Celsius). The further danger was that zirconium fuel rod casing would begin to break down at 2,200 Fahrenheit (1,200 Celsius) and melt at 3,900 Fahrenheit (2,200 Celsius).

Figuratively and quite literally, which way the wind would blow the fissionable nuclides would determine the future of the entire northwestern coasts of Japan's Tohoku and Konto regions.

NOTE

1. Staff Report, "Kan Sticks to Nuke Crisis Road Map: Meltdown Won't Halt Plan to Tame Plant by January, Leader Says," *Japan Times*, May 17, 2011, News edition.

Chapter Thirteen

Who's in Charge Here?

Tokyo Electric's myth of May 2011 stipulated that normalcy was just around the corner since temperatures in the Daiichi Reactor One flooded with water remained safe. Possibly so, but radioactive water then leaked into and through the larger, beaker-shaped containment vessel. The Nuclear Industrial and Safety Agency officials reported new data indicating that partially melted fuel had fallen to the bottom of the pressurized vessel holding the reactor core and possibly leached into the containment vessel soon after the March 11 quake and tsunami. No one knew if chunks of the fuel rods had melted through the concrete base of the containment vessel, making impossible a cold shutdown within the promised nine months.[1]

TO ESCAPE LOCAL RESPONSIBILITY AND SECURE CENTRAL CONTROL

A political drama meanwhile began. To escape the costs of the disaster and release Tokyo's traditional political controls, in May 2011 the central government began loosening its grasp over prefectures, cities, and villages. It schemed to isolate the outlying regions from the center, so that Tokyo's policies could remain unchallenged.

Towns thus had wide latitude to pursue their own ideas and recovery plans with their own people's labor and resources. Small towns like Minamisanriku near Sendai—a tsunami-obliterated former resort and fishing port now with denuded grounds and miles of rubble and rock tossed across them—drafted plans, warehoused food and water, and set up work centers and temporary housing. Some 70 percent of its buildings were ruins and of a pre-disaster population of 17,666, 1,157 people were dead or missing. The tsunami ripped away the library, hospital, and twenty-four other public facilities and left 700,000 tons of debris. It churned the ground so ferociously that it lowered the town's elevation by 2 1/2 feet. It knocked out electricity, unrestored until April 15, and the water supply, not rebuilt for six months. It killed thrity-nine town officials, including a twenty-four-year-old who stayed on lower ground, using a loudspeaker to warn of the surge. Despite the continuous and official encouragement of self-help, the destruction appalled emergency workers so much that they urged Minamisanriku's returning evacuees to reevacuate elsewhere, holding out to them the chance of returning in six months when more housing might be available. It wasn't.[2]

SHUTDOWN REFUSAL APPALLS PUBLIC

Fury beat on the door of another company, the Chubu Electric Power Company's when it refused a government request to shut down its Hamaoka nuclear plant situated just south of Tokyo atop a fault line. Chubu claimed it needed time to consider alternatives and plan protective construction, but its board met in May without reaching a decision: "The board discussed a broad range of issues including the ability to meet energy demand this summer with alternative methods, the impact on the company financials, to tsunami preparations at the nuclear plant. However, due to the potential widespread impact on our customers and shareholders, we have decided to continue our discussions at a later time." The company said it would consider the matter "swiftly"—but scheduled no follow-up meetings. A spokesman for the prime minister responded with another request for a swift shutdown, pointing out that the appeal originated from the highest levels and was unchanged. However, plant administrators had previously met government safety standards, weak as they were, so Prime Minister Kan lacked immediate authority to order a shutdown. At least his requests seemed to show his sensitivity to popular discontent. But no regard was offered to public sensitivities—and the people at risk were appalled.

HAMAOKA POLKA: BOW TO YOUR PARTNER, STEP TO THE SIDE

Before Fukushima, fifty-four nuclear plants, virtual concrete time bombs, operated in Japan, many in earthquake hazard areas, as the consequent shutdown of all but nineteen indicates. The caution this indicates came on shyly, as at Chubu Electric Power's Hamaoka 365-acre plant 100 miles southwest of Tokyo. Five reactors worked there; a sixth was coming. Safety problems closed two units before the great tsunami. Two months after the disaster, Prime Minister Kan sought an end to the entire operation; an 8.0 or higher magnitude earthquake was given an 87 percent chance to hit the area before mid-century. The government request set no time lines and noted that Hamaoka had previously met safety requirements, so Chubu politely ignored Kan.

Tokyo even helped justify inaction. The government said that absent nuclear power Japan would be energy poor. Turning off the riskiest reactors would diminish the 33 percent of national energy needs supplied, pre-Fukushima, by uranium and plutonium. Fukushima's failure meant peak-demand-season shortages. Unplugging Hamaoka would slice off some 3.6 million kilowatts of the firm's contribution of up to 30.9 million kilowatts to the national demand of 27.1 million kilowatts, said the government's Nuclear and Industrial Safety Agency. Chubu said such a loss made a quick shutdown impractical.[3]

DO-SI-DO AND BACK TO THE BAR

The government requests that shaky setups, notably Hamaoka, be halted, were touted by Tokyo as courageous steps and foresight: "the result of considering the tremendous repercussions a major accident at Hamaoka would have on the entire Japanese society." Chubu Electric there had only sand dunes to protect the elderly complex against tsunamis, and Japanese safety activists said the Hamaoka plant was only one of many vulnerable to quakes, tsunamis, or power losses from other events, but in Tokyo, as well as Hamaoka, more was said than done.

To be sure, Mr. Kan's public pronouncements came out with almost dazzling speed in a country where building consensus is preferable to debate—only two months after Fukushma was left in shambles—but then Chubu Electric took days to respond and weeks to gear up for its move toward agreement. Akihisa Mizuno, Chubu president, said the company was in agony over the projected closing of the forty-year-old reactors in the low dunes, but if so, the nation's nuclear power people were organized as well as agonized, and would show their muscle soon in financial, political, and energy policies.

The hesitant steps in Japan are in contrast to those in Germany, whose conservative government announced it would abandon nuclear power in the next decade, shuttering seventeen reactor complexes.

LITTLE LEAKAGE IN GOVERNMENT-CORPORATE CONNECTIONS

Japan's conservative and consensus-driven political culture served the needs of a nation in times of stable economic growth, but faltered in crises. "The old consensus between government and business has broken down," said optimist Michael Auslin, director of Japan studies at the American Enterprise Institute in Washington. "Everyone is looking out for their own self-interest. The old way of doing things has broken down, but there is not yet a new way to take its place." This breakdown was the product of the slow disintegration of Japan Inc.'s unity during years of national stagnation, culminating in the 2009 election victory of Mr. Kan's Democratic Party, which ended a half-century of one-party dominance.

Democrats gave lip service to challenging Japan's sclerosis, but have struggled to build and retain trust with the nation's big corporations by means such as lowering social welfare benefits and support for small business, giving tax breaks to wealthy companies and individuals, and reviving economic and political nationalism. Mistrust of the Kan government—and perhaps a desire for stronger government authority—deepened during the nuclear crisis, when Tokyo often appeared indecisive in its response and inconsistent in its release of information.

Some saw the disaster as an opportunity to shed light on the old-boy network and the government-industry revolving door. "Nuclear power plants were long treated as sacred territory," said Katsuhiko Ishibashi, a seismologist and Kobe University nuclear safety expert and agitator for tougher quake resistance standards at nuclear facilities. "The accident at Fukushima Daiichi is at last starting to change that."

Some companies went through the motions of social responsibility, as in Hamaoka, where Chubu finally decommissioned its two oldest reactors and pledged to build a seawall in the next three years to protect an area predicted to suffer an earthquake of community-destroying magnitude. Chubu said three reactors were safely operating, but agreed to "promptly consider" closing them temporarily for inspection and possible repair.

Offices and factories in eastern Japan had already dimmed lights and altered operating hours after the government request that they reduce energy use by 15 percent during the 2011 summer. "I am certain that with an even stronger energy-saving effort by both local residents and the entire Japanese nation, we will overcome the power shortage," Minister Kan said. He indicated May 7th that his government was not rethinking the nation's energy policy, telling reporters that he would not seek to close any more of Japan's nuclear reactors. He said the Hamaoka plant, was "a special case" because it sat atop a major fault line—but at least he had made the request.

THE SUMMER OF SAME OLD

Hopeful environmentalists, including Greenpeace (not the most popular group among Japan's once-powerful, pre-tsunami whaling industry) lauded Mr. Kan's decision and urged the Japanese government to consider halting operations at all reactors in Japan, atop and near shifting tectonic plates: "Greenpeace welcomes Prime Minister Kan's request to close Hamaoka, one of the most dangerous nuclear reactors in Japan," said Junichi Sato, the organization's executive director in Japan. "The government must continue to close and decommission existing plants and cancel all new reactors. Only then can the Japanese people feel their government is truly putting their safety first."

Katsuhiko Ishibashi, a seismologist at Kobe University who has long argued for more stringent seismic standards for Japan's nuclear power plants, said that simply halting operations at the Hamaoka plant was not enough. The reactors would remain vulnerable to quakes and tsunamis while they cooled down, as would the spent fuel rods stored at the plant. "Work must be carried out immediately to make sure the plant is protected," he said. He noted the glacial pace of government recognition and response to dangers to and from reactors, but he resigned when his warnings were ignored.

Scientific opinions gained more listeners and support, but oversight and corrections remained lax.

On May 17th, nuclear physicist Arnie Gundersen from Fairewinds testified that Fukushima's gaseous and liquid releases continued and even worsened: Unit 1 was in meltdown, Unit 4 leaning and facing possible collapse, several units contaminating groundwater, and area schoolchildren outside the exclusion zone had been permitted by government fiat to receive adult occupational radiation doses.

On May 13th, two months after the nuclear disaster whose effets could go on for decades, nuclear expert Gundersen pinpointed Daiichi's horrors:

- Unit 1: Inside radiation exposure remained at seventy rems an hour. That meant, in four or five hours, death.
- Unit 2: Leaking like a sieve. Seawater poured into the top came out of the hole in the bottom. The site lacked room and facilities to capture all this water, let alone spills.
- Unit 3: Hot steam blasts hitting cooler water were off-gassing a soupy radioactive fog. The heat at the top of the reactor climbed as pressure to drive water there into cooling ponds remained anemic and the ponds dried. The heat and radiation separated volatile hydrogen in what water remained, and Unit 3 consequently neared a heat explosion.

  Pictures showed evidences of earlier explosions in Unit 3's fuel pool. Other photos of the bubbling unit showed masses of shattered and crumbling concrete and metal fallen into the dried-up pool. Distortion of control rods and fuel racks indicated explosions in the pool, violent exothermic reactions carrying high levels of iodine 131. Even minimal levels of normal cooling would have dissipated the iodine-131, but after two months, there its cauldron seethed.

- Unit 4: Photographs showed its structure leaning, bent, and broken by Unit 3 fires and explosions. Unit 4 could fall in a seismic aftershock of the first earthquake, though its floor pool and fuel racks appear to have maintained their integrity; but its plutonium and other isotopes, cesium, strontium, were volatilized.

The Fukushima units used close to a hundred tons of water daily but could purify only a ton or two a day, releasing the remainder into the earth, air, groundwater, and the seas.

Asteroids of nuclear fuel and trash landed a mile from Daiichi, the Nuclear Regulatory Commission reported, meaning they would have had to be fired from the fuel pool at around 900 to 1,000 mph. Bigger pieces would be shot out at a velocity greater than the speed of sound, the big bangs outrunning the sight of the burning reactors and becoming landmarks of the new, dead, universe circling the three exploding torches the reactors had become. Unit 1 poured and seeped poisons and exploded plutonium again when technical workers tried to use nitrogen to build driving pressure inside. Unit 2 and 3 saturated, filled, clogged, and overflowed, off-site trenches now become eroding sewers. Seawater bombing meant larger outflows of nuclides, the entire site began sinking under the weight of the water and because of corrosion, building material failure, fire, and nuclear reactions.

And airborne radiation and contamination had spread in a rough circle with a forty-radius from the reactors and farther into hot spots where breathing became menaced. Within that zone, high school students donned masks and long-sleeve shirts to protect their skin; and while students were indoors, authorities were stripping out contaminated soil to reduce surroundings to levels declared safe for adults—a situation as unconscionable as that the school had been kept open at all.[4]

Other problems also emerged. In the wake of the fuel-rod meltdown in three reactors, high temperatures caused the fuel rods' zirconium covering to react with water, producing hydrogen gas. The gas built up inside the reactors, setting off explosions at two reactors and severely damaging a third. The fourth reactor not operating at the time of the accident had also been destroyed, Tokyo Electric theorizing that hydrogen may have traveled there from an adjacent reactor through pipes and vents.[5]

DOUSE THE FIRE; BUILD ON THE ASHES

The trouble at Chubu Hamaoka appeared to subside in the next weeks, but nothing had speeded up the deliberate decision-making and the reluctance of powerful companies—even in a period of crisis—to bend to directives from politicians whom they felt neglected financial issues, including contributions to political parties. They said the government's taking only three weeks to consider and issue requests for safety measures were hasty. Officials had not followed the protocol of "nemawashi," the decently paced process of consulting concerned parties to work out acceptable arrangements. Prime Minister Kan had been "undeniably abrupt," said Hiromasa Yonekura, chairman of Japan's most powerful business lobby, the Japan Business Federation, or Keidanren. "What I don't understand about the Democratic Party era is that they reach decisions just like that," Mr. Yonekura said at a news conference. "Publicizing his own opinions first, then explaining to Chubu Electric later is no more than a political performance. These decisions should have been debated first."

Hamaoka, possibly even more clearly than Daiichi, illustrated the protected status of nuclear power and big Japanese corporations generally. Because of its precarious location, Hamaoka had faced years of lawsuits seeking its suspension, but Japan's compliant courts consistently ruled in the industry's favor. Corporate Japan could run suspect plants until the heat was on, then close them and wait for objections or disagreements to dissipate. Then, as happened, they could agitate for restarts and exposure adjustments and perhaps make minimal modifications to a plant or its neighborhood.

NOTES

1. Mari Yamaguchi, Tokyo, "Japan Nuclear Reactor Damage Worse Than Expected," *Huffington Post/HuffPost World*, May 12, 2011, http://www.huffingtonpost.com/2011/05/12/japan-nuclear-reactor-damage_n_860999.html.
2. Chico Harlan, Minamisanriku, Japan, "After Japan's Tsunami, a Town Climbs Back," *The Washington Post/Collections/Tsunami*, May 10, 2011.
3. Judy Lam, "Nuclear Plant Operator Rejects Quick Shutdown Demand," *Wall Street Journal*, May 9, 2011, sec. Business, http://online.wsj.com/article/SB10001424052748703859304576308640922303086.html.
4. http://www.fairewinds.com/content/fukushima-one-step-forward-and-four-steps-back-each-unit-challenged-new-problems
5. Hiroko Tabuchi, "With Reactor Damage Thought to Be Worse, Tokyo Utility Sticks to Plan," *New York Times*, May 17, 2011.

Chapter Fourteen

Making a Nation Safer for Profits

As seen, Chubu Electric clucked over the request from Prime Minister Kan to shut and harden what was generally thought to be the plant most vulnerable to quakes and tsunamis. Its board, pressured, finally agreed to comply—with all due haste. Chubu president Akisha Mizuno called this indefinite okay "agonizing:" "The request from the prime minister is one that carries a lot of weight, and is based on growing public concern over nuclear power in the wake of the serious accident at Fukushima. We are prepared to put into place safety measures to strengthen our defenses against tsunamis. We believe this is the only way we can regain the trust of the local community and of wider society." The utility would ask the government, i.e., taxpayers, for assistance "to help avoid or lessen" the impact on its shareholders. In return, the government assured energy companies that no request would be made to close other nuclear facilities in Japan. The power companies would remain committed to nuclear energy use, the prime minister so willing.[1]

THE SPAM IN THE MACHINE

The government also put before the eyes of the nation the assurances of nuclear industry spinners that people near damaged reactors could soon return to their desolated towns, areas now and for millennia to come suffused with ionizing radiation. The government formalized this hazy promise with the title of "temporary evacuation measures," a name also promising homecomings. After all most evacuees were not much farther than twelve miles from where they had grown up, with many more in the zone twelve to twenty miles from Daiichi. In this second zone, people could plot further removals or stay indoors, wear masks, perhaps burn incense, dabble in homeopathy.

Lacking close supervision, those under voluntary house arrest, of course, went out to seek news, food, or a modicum of hope. The boundaries of the zones got little more respect from radioactive plumes, drifts, and leaks, so the neat circles they made on computer screens in Tokyo did not show them, as they were, pockmarked with hot spots.

Radiation screening itself was spotty, so inevitably the government had to deal with the issue of *cumulative* radiation exposure. Premier Kan extended the supposed no-go zone beyond twelvev miles, promised more attention to leaks in the reactor as well as the boundaries between the free-range and house arrest areas, but threw in a lagniappe for the Lower House Budget Committee: evacuees could count on being off the floors and out of the shelters by early 2012: "There may be some changes to the methods used to stabilize the reactor, but I believe we can move forward with the plan without changing the time line."[2] Japan would need their taxes.

Kan's attempt to oil troubled waters of popular discontent neglected any realistic assessment of ongoing dangers and Tokyo's trying to force local communities to shoulder the greater part of the trillion-yen (and climbing) cleanup, repair, and damage award costs. A long, hot, hungry summer began for the 80,000 people in the hills and crowded shelters 140 miles northeast of Tokyo, bereft of water, electricity, and sanitation, and the silent, ruined farms and former fishing areas along the contaminated coast.

SURVIVAL TOPS REVIVAL, YET MORE IS SAID THAN DONE

Tokyo Electric seemed newly forthright in statements on May 17; one a vow, again, to stabilize Daiichi in six months. The utility also confessed that—so sorry, no big deal—in the early hours of the crisis fuel had melted at three of the plant's six reactors, as experts had reported for months. "Although we have confirmed that the nuclear core melted soon after the accident, our three- to six-month timeline will not change," Tokyo Electric chief Sakae Muto told reporters. He said a "cold shutdown," would drop temperatures in the plutonium-uranium core to a sub-boil. The corporation would build external and structurally independent cooling systems at the crumbling reactors; flooding and filling with seawater was done.

Typhoons were expected as usual in the summer, but Tokyo Electric would wrap the shattered outer structure of Reactor One in a giant polystyrene cover to shield the building and reduce radiation emissions. Stone-filled cages, makeshift bunkers, would line the coast to protect the reactor from tidal waves. Stabilizing the reactors and the protective bunkers and polystyrene drapery would let workers start on the ten-year job of dismantling and decontaminating the site, said experts, but in a decade more geologic and oceanic turbulence might slow but not stop progress. Skeptics replied that hand-waves would not stop sea waves; they doubted even the estimate that stabilization and cleanup initiation could be done in three months.

The tons of toxins spewed and leaking from the plant—and their buildup in spots outside the twelve-mile (twenty-kilometer) evacuation zone—plus calls for action from even Kan's own party, pushed him to exile residents of more towns. The exodus was slated to take a full month, but for many that merciless month led to quickened evacuation on May 14, the rest expected to continue to move out through the end of May 2011.[3] Some of the refugees would be relocated six times, none of them ever as far from Daiichi as the 170,000 Americans ordered to get 50 miles away from ground zero.

Evacuees left behind devastation, flooding, and poisoning of their lands, although the largest hardest-hit town, Sendai, showed signs of returning to life, even re-establishing bullet train connections to the south. Smaller settlements in the northeast, however, had to worry more about survival than rebuilding. Everything they once exported was suspect: fish, produce, and even seaweed, an important crop for coastal harvesters. Livestock breeders killed their cows rather than leave them to slow starvation.

Greenpeace's radiation investigation teams verified dangerous levels in seaweed, another blow to farmers and fishermen just before the annual harvest of this vegetation and human dietary staple. The levels indicated serious problems with the entire marine ecosystem of the northeast which had provided Japan with a quarter of its seafood. Greenpeace called on the government to begin radiation testing of seaweed, seawater, and marine life along the entire Fukushima coast, inside and outside Japan's twelve-mile territorial waters. It sent samples to independent laboratories for further analysis.[4] With ocean currents carrying seaweed and irradiated fish from northeastern Japan to Pacific isles and the western coast of the United States, and with radiated clouds drifting far from Fukushima, the contamination was becoming a global issue.

Communities the tsunami ground into or off the earth and those who fled were mostly left to fend for themselves, without running water, electricity, or adequate food. Prime Minister Kan fed "keep calm" messages into the media and even made a fifty-minute visit to Fukushima, while some of his advisers hinted at grandiose steps to alter topography and create brave new settlements, removed from the sea or elevated on giant platforms or sheared-off heights, mega-projects that the government no longer could afford. Simple survival, anyway, concerned the stunned northeast much more than media grandstanding, but only wisps of information came in response to the overriding question, "Are we safe?"

Sendai became something of a showplace city, or Potemkin village, with bullet trains to Tokyo back in action a few weeks after the tsunami, shops reopening, normalcy touted far beyond the realities of smaller cut-off, dying, settlements such as Otsuchi, on the rugged north coast. Its officials frantically drafted plans to retain younger residents, needed for tax generation and civic resuscitation. It offered them little more than gratitude and unemployment in a landscape of mountains of crumpled, rusted cars, oncoming spring rains, and a hot sun baking the mess into an earthen armorplate in a flattened fishing town unable to count, much less bury, its dead, wrenched from the muck, little by little, by the Japanese Self-Defense force. "We are far from reconstruction," offered Vice Mayor Masaaki Tobai, who stepped in to lead the town and who survived by scrambling to the town hall's roof. "Medical services, administration, education, police, fire, retail stores, hotels, fishing cooperative, farming cooperative, industry, jobs—all are gone, all washed away."

Tsunami waves up to fifty feet high had torn through the town, gutted the town hall, and killed the mayor and some thirty town employees. They destroyed boats, oyster, scallop, seaweed farms, and canneries and seafood-processing plants along the wharfs. Otsuchi and other towns remained dark and destroyed into the end of the year, lacking even muscle power, since working-age people were leaving.[5] Aid was not forthcoming from corporations adamant against new taxes or from the confused, inept—some said cynical—government maze-ocrats expressing concern and advising self-help but doing little else. As summer wore on northeast Japan grew to resemble a leper colony housed in a garbage dump.

NOTES

1. Martin Fackler, "Japan Reaffirms Nuclear Energy Use," *New York Times*, May 8, 2011.
2. Staff Report, "Kan Sticks to Nuke Crisis Road Map: Meltdown Won't Halt Plan to Tame Plant by January, Leader Says," *Japan Times*, May 17, 2011.
3. "Japan to Release Revised Roadmap to Stabilize Crippled Nuclear Plant, Help Evacuees," The Associated Press/The Canadian Press, May 16, 2011.
4. Jess Miller, "Seaweed Contamination Cause for Concern in Japan," *Greenpeace Web Blogs*, May 12, 2011, http://www.greenpeace.org/international/en/news/Blogs/makingwaves/seaweed-contamination-cause-for-concern-in-ja/blog/34733/.
5. Martin Fackler, "Otsuchi. Before It Can Rebuild, Japanese Town Must Survive," *New York Times*, May 15, 2011.

Chapter Fifteen

Futures Unknown

Japan's political system is so engrained as to hinder adapatation to creeping change, let alone emergencies on a biblical scale. Too often, culturally inbuilt deference permits oligarchs to blunder or blather without backlash protest, or social strife. Moreover, the aims of corporations are clearer and more effectively pursued than public attitudes.

Inside the government-designated no-go twelve-mile ring, which includes the city of Futaba, the Environmental Ministry has pledged to reduce radiation levels by half within two years—a relatively easy goal because short-lived isotopes will deteriorate. The bigger question is how long it will take to bring levels down to what the International Commission on Radiological Protection recommends, 1 millisievert per year. That is a much more daunting task given that it will require removing cesium-137, an isotope that remains radioactive for decades.[1]

Japan's ocean moats and mountain fastnesses grant it isolation from the culture, if not economic influence, of other nations. Its wealth and order grew from unity, a well-knit fabric of cities and rural areas, an expectation of excellence in research and production, and a presumption that they, chosen people on a blessed land, can vanquish adversities—economic, technological, or natural. Thus, with the stoicism of the samurai and the solidity of the sumo, they withstand atomic war, occupation, unwieldy government, paucity of energy fuels, and a fractured economy to remain a great global power.

CONSTRUCTION PAST; RECONSTRUCTION FUTURE

Postwar, pre-tsunami, Japan's centralized economy almost erased joblessness by varying wages; by interlocked production companies passing the burden of economic downturns to smaller enterprises; by public works projects; and by maintaning swollen standing military and salaryman armies. These, the suited, were loyal to mundane lifetime appointments to sprout and nourish government and corporate offices and subsidies for projects further centralizing the economy and culture.[2],[3],[4] State-financed construction dominated the economy. Construction companies in 2011 employed 8 percent of all workers, twice the level of the U.S. Sub-contractor specialists in cement, steel, plastics, computer, and soil enhancement provided millions of workers $40,000 in annual wages, the spending of which had a nourishing trickle-up effect on the economy. Public projects and commercial and financial business employed over 90 percent of the workforce; the remainder worked in fishing, farming, food packing, or active and support duty in the military.

BMI: BUREAUCRATS MANAGING INEQUALITY

The quake and tsunami and fumbling and bumbling of the governing Liberal Democratic Party and national economic problems finally disgusted voters enough to boot the old welfare-talkers and let in the partisans for trickle-down and nationalistic economics. The not-liberal Democratic Party took over in 2009, set up Naoto Kan as prime minister, and snipped at the safety net. It drew a bead on the old WPA-U.S.-style efforts to bankroll dodgy infrastructure projects to support employment. It sliced $14 billion in 2011 from a public works budget of $88 billion as a step toward weaning regions from Tokyo spending and delivering them to the invisible invincible hand of the "Free Market," i.e., to corporations. The Kan administration also made noises suggesting a no-nukes course.

An example of outcomes in small towns was the wreck of the economy of Naganohara, a tourist hamlet north of the capital. Prosperity had been spotty and seasonal. A $5.2 billion dam construction project. the Yamba Dam, tided the town over until Kan et al. pulled the plug, as he and his Dems would do with forty-seven other public works. The advice to Naganohra was to work more and eat less, but total social disruption followed an exodus of those seeking work elsewhere, families thrown on their own meager resources, and some few calling for more aggressive entrepreneurship or even seducing a big corporation, nuclear or not, into its bailiwick.

The media and experts nourished a general attitude of "I'm all right, Jack—if you aren't, that's your lookout." More elegantly put: "What's happening at Yamba Dam is actually a revolution in the way Japan is run," said Takayoshi Igarashi, a professor of politics at Hosei University in Tokyo. "This would end Japan's structure of [local and regional] dependence on public works and central planning."[5] Many towns, though, had no bootstraps to pull up on, e.g., Naganohara with its jobless, emptiness, and shuttered businesses disagreeable to the diminishing number of tourists that found their way to it and other once-picturesque hamlets.

TSUNAMI TRUMPS, THEN CANS KAN

The tides of March 11 brought the plan to cut the Tokyo-prefect umbilical cord into squirming reality. Kan's career ebbed, too. He could cry about disaster on television, verbalize glib optimism, urge self-reliance and trust for corporations, but he simply couldn't cut the mustard: the oligarchy lamed him, inefficiencies up and down the line cost him credibility; denials, lies, and indecision made him a prime minister without a majority, or a quarterback without a helmet picking up and sending mixed signals. Or, as they say in Texas: It wasn't the heat—it was the stupidity.

The rapes of math, realities, did him. Daiichi cleanup and resttlement costs soared beyond $60 billion, a sum not including health treatments and putting the dead nuke to rest in a king-sized bed of sand, boric acid, and reinforced cement.

Confused and distracted by sidesteps, finger-pointing, and garble, the Japanese nonetheless slowly awakened to the need for another option besides the welfare state or the "free market." Safety was the main concern, but few could even find their way through the labyrinth that muddled responsibility. The Nuclear and Industrial Safety Agency, the designated regulator, hunkers within a Ministry of Economy, Trade and Industry. But its said prefects were to take care of themselves. Various ad hoc commissions toured, denied, and spun facts to support their mantra: "Fukushima, no Chernobyl." The BMI's (bureaucrats managing inequality—to use Paul Farmer's phrase in 'Mountains beyond Mountains') were right—it was no Chernobyl—it was worse, made worse by the confusion, as well, of measurement, of fluctuating designated safe levels, or competing estimates when the 150,000 refugees might regain jobs and homes and something like their former lifestyle.

Some saw through the veil. Britain's staid *Economist* said, "Fukushima revealed, again, cozy ties of the nuclear industry and government. Together, they have stifled debate, covered up bungles and made assumptions about risks that were too optimistic. The crisis management at TEPCO, the plant's owner, has laid bare an astounding lack of leadership. . . . The media, taking their cue from the Tokyo establishment, have not thought properly to report the unfolding struggle for food and fuel" throughout the northeast regions of Tohoku and Kanto.[6]

Japan Inc. would be carried through, figured its elites, on the backs of the working people. Through higher bills and taxes and the working of "the free market" these paid for the Daiichi destruction. Taxes rose for the majority (not the elites), higher electrical charges eased the burden on utility companies' token reparations, while shortages and market prices crept up. Millions outside Tokyo took up post-apocalyptic burdens. Towns tried to restructure and rebuild with local resources while disorderly caravans of the viable labor force sought economically viable regions. The nation aged; the birth rate fell.

NOT THE LAST TYPHOON

The new austerity, partly policy, partly disaster, partly indecision, threatened—and threatens—Japan's future. With nuclear concerns and problems ongoing, Prime Minister Kan said that Japan, the world's third largest economy, would end plans to double its nuclear power capacity and build new nuclear plants. Japan would "start from scratch" with a new energy policy by developing wind, solar, tidal, and biomass energy sources. Kan, however, helped restart several of the old plants—after more or less thorough safety checks. At any rate up and running again. But he was gone by September, replaced by a former finance minister, a war crime and comfort woman denier, maybe with a few new moves but still dancing to the profit-takers' beat.

Kan's administration also lost credibility also because of his crumbly stonewalling. For example, Tokyo first denied meltdowns ("No Chernobyl here!"). Then three months later, given mounting evidence and probably international disclosures, Kan's team finally—in May—admitted to not one but three Daiichi meltdowns, all known and covered up within less than a week after the earthquake.[7]

Tokyo made some safeguarding motions. That last week in May, it pulled the switch at Hamaoka Nuclear, an old power producer open to earthquakes and tsunamis. Operators of Hamaoka grudgingly, slowly, complied, and it came out that one reason it had not earlier been put out to pasture was its big buys of British MOX fuel, which blew things apart at Fukushma. Other plant closures and popular resistance to nuclear power continued.[8]

Even this environmental awareness of the free enterprisers came long, and slowly, after the May 11 disaster. Meanwhile radiation poisoning spread. By May 27, soil in a 600-square-kilometer area, mostly to the northwest of the Fukushima plant, indicated absorption of radioactive cesium of over 1.48 million becquerels per square meter, the yardstick for compulsory evacuation orders in the 1986 Chernobyl catastrophe. And more poison was blowing in the wind. Strong breezes wafted much, but not all, radiation out to sea and as far as California. Contaminated soil from Daiichi whirled across a land area larger than the Chernobyl fallout toxified.[9] Other countries began to back away from nuclear reactor plans, nuclear power companies tanked in the stock markets, or as one analyst put it investors seemed to have "lost their nerve before their shirts."[10,11] But investors saw that, even in tech-savvy

Japan, nuclear plants risked technical failure, human error, and natural disaster, or combinations thereof.

By the end of May 2011, almost four months after the meltdown and explosions and leaks, evacuations, and fumbling, polls showed that four-fifths of the nation distrusted government information on the crisis.[12] Financial storms buffeted Tokyo Electric, as its stock price fell on a loss of 1.25 billion yen in March, the biggest loss outside the financial sector in Japan's corporate history. In fiscal 2010 it had reaped a profit of 134 billion, but the new problems hindered its applications for credit for rebuilding.

Its government-backed bonds brought little help, since the government itself was expected to go deeper into debt funding reconstruction and reassurance efforts. Japan's national debt relative to the size of its economy already towered over that of other advanced economies, and greater debt would limit Tokyo's ability to help utilities. Thus, both Japanese and world bankers downgraded government bond ratings, slowing reconstruction efforts and weakening political oversight from Tokyo.[13] And with 97 percent of the debt held by Japanese financial institutions, both they and global investors would risk buying new debt only if interest rates were increased above the then existing 1.1 percent. Not wanting to enlarge annual outlays to pay such rates, Tokyo explored an alternative: taxing the Japanese population, even as it cut public employment.

RUNNING AWAY FROM JAPAN'S CRISIS

Japan's corporations worried that the nuclear crisis would lead to decades of economic turmoil and shrinkage of home markets. The yen was strong; so they bought, or bought into, production abroad rather than pursuing symbiotic relationships or investing at home to create more competitive enterprises. The outsourced economy, of course, hurt the job rate, rebuilding, and the marginally employed. From April to June, Japanese companies spent $26.6 billion on mergers and acquisitions overseas, the highest quarterly volume in almost three years.

Not every corporation could flee overseas. Sony, with employment contracts, has long had its hands tied in trimming its global work force of almost 170,000. A third of these are in Japan, where large-scale layoffs by top companies are highly controversial. Sony's insistence on making many of its own televisions and other gadgets in high-cost Japan, instead of outsourcing to less-expensive manufacturers, was a strategy the company continued. Despite the gadgety surge, Sony still regards "TVs as a core Sony business," said Kazuo Hirai, an executive vice president. "It is Sony's DNA," he continued. "If we outsource everything overseas, we will lose that."[14]

Some tried to look on the bright side of the strong-yen situation. "When the yen is on an upward trend vis-à-vis other currencies, it creates a favorable environment for mergers and acquisitions," spun Kotaro Masuda of the Institute for International Trade and Investment. "It also becomes possible for companies to buffer against currency risks by producing more overseas."

More experts saw an easier way to profits and increased exports: cheapening the yen. Though companies were trying to rebuild factories and restore supply chains, the strong yen made exports pricey overseas. Joint ventures, motor awards, and lower prices made non-Japanese products such as Hyundai and KIA autos better sellers than the former stalwarts of prosperity such as Honda, Toyota, and Sony. Honda Motor Company tearfully announced a net income fall of 56 percent to 60.4 billion yen, or $796 million, in its 2011 fiscal second quarter, compared with 135.9 billion yen a year before. With the yen trading at an average of 78 to the dollar, compared with 86 a year earlier, Honda said exchange rates knocked 36.7 billion yen off of its books. Panasonic Corporation, on October 31, 2011, blamed the strong yen for killing profits—it had forecast a profit of 30 billion yen for 2011 but instead announced a 420-billion-yen loss in the twelve months through March.[15]

The weather turned foul on the Nikkei stock exchange and in supply and assembly areas serving the corporate giants. Sustained midsummer flooding in Thailand hurt part-making plants, whose lowered output in turn hindered U.S. assembly plants and forced Honda to cut North American factory production by half in November. Toyota, after the Thai floods, largely ended overtime in North America in late October, 2011. U.S. and Thai workers were laid off en masse.

Aside from pressing for a cheaper yen, some corporations tried to move their entire operations and profit harvesting to low-wage areas abroad. They soon surpassed their 2008 record of $68 billion in foreign investment and acquisition.[16] Many companies thought it would take too long anyway to resume domestic operations in a country with disrupted supply chains, carefully designed planning and profit projections trashed, and many facilities in rubble. The tactic of just-in-time deliveries of necessities to manufacturers and products to wary consumers became impossible, even when low-wage, non-union workers, not much more than temporaries, would docilely accept speed-ups, occasional longer or shorter hours, or other corporate disciplinary methods.[17,18]

Corporations handled pretty well the problem that tainted or suspected food might require the import of safe farm products and capital export. The big companies counted on fear permitting higher prices and profits. On September 6, Kansai Super Market Ltd. announced a 30 percent stake in Masami Cattle Ranch Inc. in California which would expand production and supply Kansai stores in Japan.

A NAFTA-like scheme was only partly successful. Japan, the U.S., and Australia hammered out a Trans-Pacific free trade agreement in late 2011. It eliminated tariffs on Japanese exports, but also required reduction of restrictions and fees on imports. Japan's rice farmers had long been protected under sky-high tariffs, such as a 777.7 percent levy on imported rice. In a powerful campaign against the pact, farmers portrayed the tariff removal as a threat to the culture and agriculture of Japan. To assertions from proponents of free trade that consumers would benefit from cheaper imports of goods and food, for which they pay some of the highest prices in the world, domestic producers responded that their enterprises and Japan's rural traditions would be ruined by foreign mass production making Japan import-dependent.[19]

RADIATION CHECKERS

An Agriculture Ministry report, however, boosted demand for foreign foods in August just before the rice harvest, previously most productive in Japan's northeast. The Ministry had found higher than trace levels of cesium in rice but denied any health risk. Inspectors in Ibaraki Prefecture, just north of Tokyo, also found radioactive cesium in a rice sample from Hokota city, about one hundred miles south of the radiation geysers. The prefecture said the radiation was safe, fifty-two becquerels per kilogram, about 10 percent of the government-set limit.

Radiation at unsafe levels had already been found in beef, spinach, and green tea, and shipment stopped from some prefectures. Cesium in the beef of Fukushima Prefecture put it on the embargo list. But the ministry said it would lift a similar ban on beef from Miyagi Prefecture, which borders Fukushima to the north, after Miyagi farmers took measures to limit radiation exposure to cows, by feeding them rice straw from areas away from the northeast.[20] But who wanted meat, rice, tea, vegetables, and seafood, even from a "safe" prefecture next to a toxic one? The official safety level also came into question, as indicated by the comment of Tatsuhiko Kodama, head of the Radioisotope Center at Tokyo University: "There is no safe level of internal radiation exposure, especially for children."

The establishment of the twelve-mile evacuation no-farm zone reassured few. Farming ceased but dangerous radiation levels were turning up in tea, milk, fish, beef, and other foods produced as far as two hundred miles from the plant! And radio-checks turned out to be spot, then spotty: prefect officials simply lacked the capacity to screen all farm produce in the north for radiation. Moreover checking and zoning could be evaded: in July the media let the public know that more than one thousand cows had been shipped to market and slaughtered after being fed radioactive straw. The meat was tracked to supermarkets and school lunchrooms and the stomachs of thousands of consumers, including schoolchildren.

The health ministry in August found hay contaminated with as much as 690,000 becquerels a kilogram, compared with a government safety standard of 300 becquerels, in cattle feed. The government instructed farmers to stop feeding the contaminated hay to cows and promised to test all meat from farms storing hay outdoors. "We have now set up safety control measures monitoring feed and testing in each prefecture," Chief Cabinet Secretary Yukio Edano told reporters. But, again, the process was unwieldy, the job too big, so only meat from one cow per farm got a going-over. And hurried radiation checks with crude, handheld dosimeters simply did not show when cows had eaten cesium-laced hay. Given the failure of the screening program, Edano in mid-August lifted the meat ban for Miyagi Prefecture, then for Fukushima, Iwate, and Tochigi Prefectures in eastern Japan—a step delayed six days by the discovery that a Fukushima farmer had shipped more contaminated beef. The radiated beef ban and inspection program had lasted one month.

The very wood from the tiny remnants of Japanese forests turned suspect. The national forestry agency said contaminated tree products from Fukushima, like farm products from the northeast, were everywhere. On September 6th, the farm ministry ordered local authorities to prevent shipments of any wood or charcoal that had been stored outdoors since the nuclear crisis and, particularly, to test on trees used for mushroom growing. Nameko mushrooms grown outdoors in Soma, a city about twenty-five miles (forty km) north of the Daiichi plant showed nine times the legal limit of cesium, the local government had already stated in early August. The farm ministry asked growers in Fukushima prefecture to stop harvesting mushrooms off raw wood left outside,[21] but mushrooms remained on school lunch menus, said the ministry, and would continue to be served as part of the obligatory student diet. So the mushroom farmers continued to profit while children ate their bitter fruit.

The government, the media, and a Tokyo University study agreed that Japan's wheat would escape impact from the nuclear disaster because cesium levels in the roots of the plants are low, and the effect on the wheat spikes for consumption were probably safe. But few trusted anything the government and its agents said, especially as it impacted children.

The distrust and anger grew when traces of radioactive cesium thought to be from Fukushima were detected in Japanese baby formula in December, nine months after the accident. Meiji, the Tokyo maker of the powdered formula, announced the recall of 400,000 cans of formula, but said the recall was only a precaution since levels detected were well below the government's safety limits. Tests found a combined 30.8 becquerels per kilogram of cesium-134 and cesium-137, the company said, compared with the government limit of 200. (A becquerel is a frequently used measurement of radiation.) A Meiji spokesman, however, admitted airborne cesium particles might have contminated formula at a factory in Saitama, north of Tokyo. He said the company had diligently screened radiation levels in the water, but had taken less care to filter for airborne radioactivity. Supporting Mejii, Eriko Matsuda, of the government-affiliated National Institute of Health Sciences, told public broadcaster NHK that formula was made safe by dilution in water before use—but once again—as with farm goods—quantities of product went unchecked, some shipments were only partially checked, and consumers were hardly reassured.[22]

Radioactive iodine already had been detected in the thyroids of half of 1,000 Fukushima children, broadcaster NHK reported September 6th, citing findings from a group led by Satoshi Tashiro, a professor at Hiroshima University. Tashiro said children should continue to be monitored though, as usual, the levels were not admitted to be dangerous.[23] The declarations of safety and calls for calm seemed a government-corporate tactic to minimize the harm to farmers from the Fukushima area and bring back local economies, and the tactic was used often, even to prevent incipient chaos among refugees and evacuees. Yet reports of the seeming omnipresence of long-lasting radioactive cesium accumulated, undermining public trust in the safety of anything from or near the region and the validity of the calls for calm.[24]

In a fashion the calm message was a way of blaming the victims: panic, anger, or protest meant one was going against the cultural grain of which the police and military grew ever more solicitous. An egregious example of such blaming came about when Tokyo Electric argued that people or organizations on whose property radioactive materials had fallen or could be detected were responsible for cleanup and costs.[25] The utility, in court, said it was not responsible for decontamination of a radioactive golf course because it no longer "owned" the radioactive substances: "Radioactive materials (such as cesium) that scattered and fell from the Fukushima No. 1 nuclear plant belong to individual landowners there, not TEPCO," the utility said. The court and an appeals court did not accept that argument, but assigned no blame or costs to TEPCO with a decisions that cleanup was the responsibility of the local and central governments. The two golf companies with radiated greenways had filed for a provisional disposition with the Tokyo District Court, demanding TEPCO decontaminate the golf course and pay about $1.13 million (87 million yen) for upkeep for six months. The sum, if it became a government resposibility, would slash budgets for many communities northwest of Tokyo.

The golf course has been out of operation since March 12, the day after the Great East Japan Earthquake and tsunami set off the nuclear crisis. The companies wanted to reopen the course in July, but radiation levels, checked by the Nihonmatsu municipal government in June, were above the national safety limits. On Aug. 10, a level of 2.91 microsieverts per hour was recorded, 10 centimeters above ground at the tee of the sixth hole. The level was 51.1 microsieverts per hour near a drainage ditch in a parking space for golf carts, a level comparable to the Ottozawa area of Okuma, 2.4 km from the plant. But Tokyo Electric challenged these figures: "There is room for doubt about the ability of the measuring equipment the city used and the accuracy of the records." And even if the numbers were true, there was nothing to worry about: "There are sites overseas with an annual reading of 10 millisieverts of natural radiation."

The case was dismissed by the appeals court in November; the golf course owners appealed to the Tokyo High Court, and the courses have remained closed until a definitive ruling is issued and funds found for neutralization of the grounds. On Nov. 17, radioactive strontium at ninety-eight becquerels per kilogram was detected in the grass and ground. Asked about TEPCO's doubts concerning the city's radiation measurements, Nihonmatsu Mayor Keiichi Miho said, "We made the utmost efforts when we conducted the checks."[26] Meanwhile one wag, Rob Harris, put out a fake resolution story on the Web's Permalink site: "Night Golf Available—Bring Lead Shield."

RUBBLE TROUBLE

Where, though, to put the rubble in the aftermath of the great tsunami?

A recycling industry to clear away 25 million tons of tsunami-smashed cars, homes, roads, bridges, and other structures—wood, steel, concrete, and plastic—would have to be directed and coordinated by prefectural authorities, with assists from Tokyo. Recycling, financed by the central government, some estimated, could create higher employment and put Japan on the road to rehabilitation in less than a decade.[27]

Rather than moving recycling to the front burner, though, Tokyo decided to mother-hen the surviving functioning nuclear plants. They that, like Daiichi, use eight-hour batteries were now required in case future nuclear-power-plant generators failed to stop cooling fuel from overheating—leading to meltdowns. At Daiichi, these backup batteries clearly were not sufficient to avoid meltdown.[28]

Nuclear power will remain at the eye of Japan's future political storm. In his moment of weakness, Prime Minister Kan failed to persuade the public that his new antinuclear stance represented a serious change of heart by a leader who before March 2011 supported building new nuclear plants. He failed to lay out a detailed road map of how Japan might realistically rid itself of nuclear power without industry operating half-time and all the lights going out. At the end of August, the government announced that it would halve radiation levels in areas contaminated by the stricken Fukushima Daiichi plant within two years. But it failed to explain how it would do so, or how it would pay the considerable costs.[29]

The present as history is upon us. Great debate and uncertainty still beset the nation over the closing of Nippon's nuclear power plants. Daiichi should be dismantled forever. Perhaps others will be shuttered for repairs and rehabilitation. But those providing the foundation for the remaing 25 to 33 percent of Japan's electrical energy might still be dangerously maintained—until the next nuclear tsunami comes in the wake of a massive earthquake—even a millennial last wave.[30]

Under Tokyo Electric Power Co.'s road map to deal with its crippled nuclear plant, reducing temperatures at the bottom of the core pressure vessel is one objective for bringing the accident under control. But if the fuel burned through the pressure vessel surrounding the No. 3 reactor and dropped into the containment vessel, that plan would be affected.[31]

Prime Minister Noda thus faces continuing safety and, especially, financial issues. To fund recovery, he must implement what amounts to a separate national budget to come up with a sum of $169 billion over the next five years. A fiscal conservative, Noda suggests that raising taxes might rein in Japan's deficit. Even more so, others in his party argue that economic weakness precludes further burdening consumers and corporations. Presently, tax income covers less than half of Japan's annual outlays, but the Noda administration cannot escape post-quake costs, put at $260 billion (20 trillion yen) for the recent disasters, excluding costs from the continuing Fukushima crisis.

That means spending on a scale unprecedented even for Japan, whose decades of public works projects pushed the national debt to 228 percent of its gross domestic product in 2011. That far exceeds even debt-saddled Italy, at 118 percent, and the United States, at 92 percent, according to the International Monetary Fund. And in trying to raise taxes, Mr. Noda must contend with a painful historical precedent: an increase in the consumption tax meant to help finance rebuilding after the 1995 earthquake in Kobe snuffed out a nascent economic recovery. The consumption tax has not been raised since.[32]

Dickering and dithering resulted in less accomplishment than Nero's famous fiddling. "There has been so much delay over the past months" Nobuteru Ishihara, the secretary-general of the Liberal Democratic Party, told the public broadcaster NHK, "Mr. Noda must make up for that delay." But the Japanese Diet was in gridlock, the opposition parties controlling the legislature's upper chamber to block Democratic Party initiatives. The opposition can cause delay and blame it on the muddled Noda administration for inaction.[33]

Meanwhile, nuclear isotopes spread across political boundaries and international agencies and personages wring their hands over the effect on the Japanese, especially children:

The three co-presidents of the Nobel Peace Prize laureate International Physicians for the Prevention of Nuclear War (IPPNW)—Sergey Kolesnikov of Russia, Vappu Taipale of Finland, and Robert Mtonga of Zambia—wrote Japanese Prime Minister Naoto Kan on August 22nd urging immediate action to better protect those downwind and downstream of Fukushima Daiichi's radioactive fallout. Among its many urgent calls, IPPNW urges that preparations be made to evacuate residents as far as one hundred kilometers (as opposed to the current twenty kilometers) from the three melted-down atomic reactors, if radiological conditions demand it. IPPNW also calls for a return to the pre-catastrophe 100 millirem-per-year dose limits for the population, instead of the controversial twenty-fold increase (to 2 rem per year) in "allowable" radiation doses for Fukushima Prefecture residents established by the Japanese federal government in April.[34]

Radiation went to sea and air and fell on hot spots on land. Toshimasa Ohara and his colleagues at the National Institute for Environmental Studies in Tsukuba, Japan, simulated the transport and deposition of iodine-131 and cesium-137, two radioactive materials released in large quantities. They found that 13 percent of iodine-131 and 22 percent of cesium-137 fell on land. The deposition created zones in which cesium-137 levels reached up to 20 kilobecquerels per square meter as far away as the northern part of the Kanto Plain, some 250 kilometers from the plant.[35]

Promises, Promises

The use of nuclear energy continues, despite much "on the land" hand-waving. Prime Minister Oda's inaugural address called for phasing out nuclear power, but stressed that reducing Japan's dependence on nuclear power could only be gradual. Reactors idled over safety fears after Fukushima Daiichi accident would get complete physical and functional exams—then restarted, along with a process of education for host communities. Said Noda, "To build new reactors is unrealistic, and we will decommission reactors at the end of their life spans. But it is also impossible to immediately reduce our dependence to zero," he added.[36]

TEPCO, Financial Bully

TEPCO is a financial bully, its enormous domestic influence rivaling the American defense industry. Its virtual monopoly and secretly calibrated electricity pricing system has empowered it as a great cash machine supporting politicians, bureaucrats, and businessmen, who have repaid TEPCO with grovelling support.

TEPCO—whose net income was $1.7 billion in 2009—has a vast apparatus for generating wealth that includes 192 plants powering a third of Japan, as well as an ancillary network of related companies to which it hands out inflated contracts, while some of those companies, in turn, arrange deals with large manufacturers, allowing them a share of the wealth.

TEPCO is here to stay, too, forcing utility customers to pay higher rates as well as bail it out with the help of other utilities' compliant contributions during the post-Daiichi era. There will be no reform, no deregulation questioning TEPCO, that would separate its electrical power production from its distribution to customers paying twice the rates charged in the U.S. And Japan is unlikely to escape TEPCO's future nuclear hammerlock on the nation, awaiting perhaps another nuclear tsunami.[37]

Japan could, or could have, become the first nation to deploy a labor force to capture the sun's energy, draw on sea winds and tides, and reprocess biomass to generate six times the electrical energy it once produced using nuclear power to boil water to create stream to turn turbines. Today's "Marshall System" also claims to be a unique way to harness the deep-ocean hydrothermal vents for energy, mining, and water desalination. It is clean and green. In contrast to the largest U.S. nuclear plant supplying energy enough for 4,000,000 homes, a single Marshall System plant could power 20,000,000 homes. Vents are not hard to find in the Sea of Japan and the Pacific, and their use could permit scrapping other polluting and dangerous technologies.

WHO THE HELL'S IN CHARGE?

Advice and commentary from outside Japan range from undoable science to unacceptable criticism. Much of it has been hindsight, as was the rhetorical question from Britain's *Economist*: "Who the hell's in charge?"—offering its own answer:

"Japan has gone without effective leadership for so long, with an endless procession of faceless prime ministers and their cabinets that it has made political dysfunction look almost like a well-practiced art. But this crisis has shown a pitiless light on that failiure. Mr. Kan, who has promised political change, now needs to bring it about. Japan's people can help, adopting a different attitude to their government. Stoicism—however good for coping with adversity—is bad for bringing on change. Time for the Japanese to unleash some righteous anger on a system that has let them down."[38]

The Japanese are angry, yes. The leaks of Fukushima have passed the status of leaks to the media and become splashy lead stories. Hiroshima and

Nagasaki rememberers join those from the hurt provinces and anxious consumers everywhere in opposing nuclear electricity. Reporters submit stories about the rising popular anger to local editors; but the editors sit on them, afraid of political and economic reprisals from advertisers, prefectural governors, and Tokyo authorities doling out funds for post-tsunami rehabilitation.

An example perhaps unequalled took place years ago in 1982, on Iwaishima, an island of about 1,000 people two and a half miles from the land of the Chugoku Electric Power Company site. It planned building a nuclear power plant on a deserted tip.

At first residents greeted the news. Chugoku Electric wooed them, paying for lavish "study tours" to nuclear reactors around the country—trips that included stops at hot spring resorts. It also offered local fishing cooperatives compensation for the loss of fishing grounds to be filled in to build the 3.5-million-square-foot plant. "The town needed the money," said Katsumi Inoue, sixty-seven, who led a pro-nuke movement on the island: "We needed to grow." But the island's fishing cooperative voted overwhelmingly against the plans. On a chilly morning in January 1983, almost four hundred islanders cut short their New Year's festivities to stage a protest march, the men in their fishing boots and the women in bonnets, through stone-walled alleys. More than one thousand protests followed in the next twenty-five years, yet the plant may still be built. Since the Daiichi meltdown, however, the protest has intensified: "We are going to stop them completely," Sadao Yamato, a leader of Iwaishima's protest movement, said at a recent rally. "It's the best chance we've had in three decades."[39]

Crusades, rural, popular, media, political now or in the future, however, cannot make the past truly past. The last lethal rays will not subside for ten millennia. The wide dispersal of their vectors imperils the very survival of Japanese and Japan. One hopes for a Japanese spring, with renewal and reinvention, rebuilding and rebellion against familial oligarchies and their lust for lethal nuclear plants and against U.S. and French pressures to retread the terrible schemes of Cold War years such as Atoms for Peace—a peace to end all peaces. And many experts fear that the government is declaring victory at the Fukushima Daiichi nuclear plant only to appease public anger and could deflect attention from remaining radiation threats.[40]

WARNINGS FROM THE SEA

Nature may be unpredictable but it is not dithering. Its motto might be taken from the Pinkerton Detective Agency: "We Never Sleep." It shows its hand to those able to see, as on September 4, 2011, in typhoon Talas, the season's twelfth. Overnight, Talas crossed the southern island of Shikoku and central

Honshu, Japan's main island. It crawled north in the Sea of Japan, off the country's western coast. Japan's Meteorological Agency said its record rain in western and central Japan, killed at least twenty-five people and stranded thousands, turned towns into lakes, washed away cars, and set off mudslides that buried or destroyed houses. At least fifty people were missing. Evacuation orders and advisories were issued to 460,000 people in the region, but flooded rivers, landslides, and collapsed bridges prevented escape and rescues. The storm moved slowly north, and with the ground already soaked, fears of additional mudslides were growing; and the agency issued landslide warnings in nearly all of the country's prefectures.[41] And as this storm washed across the country, many feared it would spread lethal nuclides all across the land and sea, even beyond Japan.

The greatest fear remained the continuing radioactive leaks at Daiichi. Since the disaster in March, workers had been struggling to cool the stricken plant's reactors by flooding them with water, which was then contaminated with radioactivity in the process, multiplying water-contamination. TEPCO installed a new circulatory cooling system in September with filters that decontaminate and recycle the cooling water. But the company acknowledges that some water has already leaked into the ocean, and thousands of tons of water remain in the flooded basements of the plant's reactor buildings.

TEPCO also admitted on December 4 that new water leakage had about one million times as much radioactive strontium as the maximum safe level set by the government, but appeared to have already been cleansed of radioactive cesium before leaking out. Both elements are readily absorbed by living tissue and can greatly increase the risk of developing cancer.

The Institute for Radiological Protection and Nuclear Safety in France estimates that between March and mid-July, the amount of radioactive cesium-137 that had leaked into the Pacific from the Fukushima Daiichi plant amounted to 27.1 peta-becquerels, the greatest amount known to have been released from a single episode. (A becquerel is a frequently used measure of radiation, and a peta-becquerel is a million billion becquerels.)[42] By early December, 2011, at least forty-five tons of highly radioactive water had leaked from a purification facility at the Fukushima Daiichi nuclear power station, and some of it may have reached the Pacific Ocean. It will continue to leak too.

For the most technically challenging step is yet to come—removing the melted fuel, a process that the government said will take twenty-five years and require new types of robots and other new technologies to remove melted fuel that appears to be stuck to the bottom of the reactors' containment vessels. These technologies have not even been developed yet. And after the removal, fully decommissioning the reactors will take another five to ten years, according to the roadmap—so fully shutting down the three wrecked reactors at the Fukushima Daiichi nuclear plant likely will take forty years.[43]

SKULLDUGGERY AND THE LYING GOVERNOR

To worsen matters, the nuclear power brokers have resorted to dirty tricks to try to quell public revulsion against the nuclear energy. On September 8, 2011, an independent investigative committee said a nuclear plant operator had tried to manipulate public opinion with fake e-mail. The operator abetted a top local government official, aiming to fire up again nuclear reactors idled after Fukushima.

The manipulation involved Yasushi Furukawa, governor of Saga Prefecture at the southeastern tip of Japan. He directed Kyushu Electric Power to send pro-restart e-mails. These were made to seem the product of average citizens calling for the resurrection of two reactors at the company's Genkai Nuclear Power Station. The company had been caught earlier ordering employees, posing as regular citizens, to send e-mails during a June online town-hall-style meeting over whether to allow the restart of the reactors. Despite the company's admission, the investigating committee did not in either case accuse the governor of explicitly asking officials to send e-mails masquerading as coming from the public, but only of asking it to send e-mails.

Tryng to save face, Furukawa insisted a Kyushu Electric vice-president had misheard his private remarks. The public scoffed at the cover-up; anger surged. Money had changed hands, and Furukawa was a long-term loyal Kyushu bunkmate. The scandal went national because if Furukawa backed off from pushing for restart, his decision could sway leaders elsewhere facing the issue of restarting their local reactors. If they turned thumbs down, the nuclear power industry, in bad odor with its payoffs and tricks, might have to turn off its lights. Its deeds spoke louder than its claim that it provided a third of the nation's electrical generating capacity, jobs, and export capabilities. Despite more evident public opposition and skepticism, most strongly believe officials go on working secretly with and for the nuclear power companies.[44]

By mid-September some forty-three of Japan's fifty-four reactors sat idle, few because of March 11th damage, most for maintenance checks legally required every thirteen months. To show some semblance of responsiveness to public concern, Tokyo asked local governors and mayors to sign off before any undamaged reactors went back online. Should none be restarted, Japan's last operating reactor would shut down by April 2012. But energy companies are nothing if not energetic—and Japanese politics have not always exemplified governance of, by, and for the people; and infant alternative energy projects may be strangled in the cradle—so the struggle continues.

Indeed, there is no end in sight for the smoldering Daiichi nuclear plants. On November 2nd nuclear workers at the crippled power plant raced to inject boric acid into the plant's No. 2 reactor after telltale radioactive elements were detected there; and the plant's owner admitted for the first time that fuel deep inside three stricken reactors was probably continuing to experience bursts of fission.

Again the technical sphere would trump and endanger the biosphere that sustains food production and the safety of the population. TEPCO and the nuclear scientists had repeatedly lied and tried to confuse the public by using technical jargon and purported facts. Few people could understand them, but well knew they remained in danger of being irradiated.

In the jumble of material conditions in the damaged reactors, the scientists insisted that detected radioactive elements *seemed* very unlikely to be able to produce *sustained* fission. But some experts had long suspected *intermittent* criticalities releasing nuclides without causing an explosion. In other accidents, nuclear material had burst into fission, but the released energy then rearranged the damaged fuel into a configuration that no longer supported fission. Yet, very gradually, the material re-formed in a way to support *another burst.*[45]

The meanest blow might yet come. Molten nuclear fuel may have bored into the floor of at least one of the reactors at the stricken Fukushima Daiichi nuclear power plant, the complex's operator revealed on November 30, 2011, citing a new simulation of the accident that crippled the plant in March.

The simulation suggested that the meltdown may have been more severe than had previously been thought. Soon after an earthquake and a tsunami on March 11 knocked out cooling systems at the power plant, nuclear fuel rods in three of its six reactors overheated and slumped. In the No. 1 reactor, the overheated fuel may have eroded the primary containment vessel's thick concrete floor, and it may have gotten almost within a foot of a crucial steel barrier, the utility said the new simulation suggested. Beneath that steel layer is a concrete basement, which is the last barrier before the fuel would have begun to penetrate the earth.

Some nuclear experts have warned that water from a makeshift cooling system now in place at the plant may not be able to properly cool any nuclear fuel that may have seeped into the concrete. The new simulation may call into question the efforts to cool and stabilize the reactor, but the Tokyo Electric Power Company, or TEPCO, says it is not worried more than eight months after the accident.

The findings are the latest in a series of increasingly grave scenarios presented by TEPCO about the state of the reactors. The company initially insisted that there was no breach at any of the three most-damaged reactors;

it later said that there might have been a breach, but that most of the nuclear fuel had remained within the containment vessels. "This is still an overly optimistic simulation," said Hiroaki Koide, an assistant professor of physics at the Kyoto University Research Reactor Institute, who has been a vocal critic of TEPCO's lack of disclosure of details of the disaster. TEPCO would very much like to say that the outermost containment is not completely compromised and that the meltdown stopped before the outer steel barrier, he said, "but even by their own simulation, it's very borderline."[46]

So the safety of the public turned and still turns on probabilities and guesswork, meaning all bets are off. No one really knows when nuclear fission will lead to another explosion and burst the geographic twelve-mile safety zone the Japanese authorities proclaim is safe for the population to occupy. Exodus is their only real hope—but where will they go in such a small chain of Japan's wondrous islands?

DOUBLE-THINK; DOUBLE-BLINK: LOOKING BEYOND JAPAN

The West, especially countries complicit in the catastrophe, is rethinking dependence on nuclear power. The U.S. Nuclear Regulatory Commission has extensively reviewed the nation's 104 nuclear power reactors and found evidence that the dangers of nuclear power have been underestimated in the past. An Associated Press analysis of a preliminary government report on nuclear reactor safety put the risk of an earthquake plus accident at a U.S. nuclear plant at twenty-four times greater than previously thought, suggesting an urgent need for upgrades.

But some opponents of nuclear power find no amount of planning or patching sufficient. Risks to the environment and human health, they say, overshadow putative nuclear power benefits—even if those include decreasing fossil fuel use.[47]

Daiichi's may have been the explosions heard around the world, even in China and India's new nuclear sites, near British and French power fuel cauldrons, and offices once issuing pieties about Atoms for Peace. Still, the UK, France, and the U.S. have little in the way of anti-nuke activism. Japan has more, but its effect is uncertain since Prime Minister Noda reveals more openly support for "limited' nuclear plant operation—and Japanese protestors roar like lions, with their protest chants and marches, but act like lambs in their actions: petitioning a pro-nuke administration to change its ways.

Over 50,000 protestors, for example, turned out in mid-September in Tokyo for a Goodbye Nuclear Plants Rally, supported and attended by popular musicians, Nobel Prize winner author Kenzaburo Oe, liberal and leftist political groups, trade unions, and Fukushimans. One of these, Ruiko Muto, told the crowd from the stage, "We now know that the facts about the crisis are not revealed, the government does not protect the people, the Fukushima accident is still ongoing. . . . But there are people who still promote nuclear power."

Rally organizers said they will collect signatures of ten million people who agree to ending nuclear power dependence by halting construction of atom-power facilities and decommissioning existing plants in a planned manner. They said they have about one million signatures so far. They will continue collecting signatures until next March and submit them to the Diet and Prime Minister Noda. A cynic might ask, what are ten million signatures to ten trillion yen? Or which is greater in Japan: fear of radiation or fear of China? Or would it not require a thorough revamping of the Japanese economy, the mind-set of oligarchs, and policy changes even to reawaken basic eco-consciousness in Nippon—or elsewhere? This or that crisis du jour sets all the Chicken Littles, sincere though they be, cackling, while the instigators and perpetrators of nuclear plants walk free in the shadowy alleys of Nippon-style capitalism, keeping their political switchblades out of sight.[48]

SHELL GAME

A great shell game is going on at Fukushima, with massive irradiated water releases into the ocean being counted as zero as the year 2012 begins. TEPCO and Japan's government prepare for massive releases of radioactive water into the ocean threatening world health.

The illogic of the zero computation goes like this, according to *Tokyo Shimbun's* interview with NISA (Nuclear and Industrial Safety Agency). The agency has regarded the radioactive water leaks from the Fukushima Daiichi reactors as "zero" in their legal term, for the reason that it is a "state of emergency." They plan to continue to regard any future leak or intentional release as "zero." The Japanese government plans to announce "cold shutdown" on December 16, but it is a questionable attitude for the government to neglect one of their criteria of "cold shutdown," which is "radioactive release being under control."

According to law, electric companies are required to set the maximum amount of total radioactive release into the ocean per year for each nuclear power plant. That amount for Fukushima Daiichi is 220 billion becquerels (cesiums, etc.), and when the year changes, they reset to "zero."

On April 2, highly radioactive water was found leaking from Reactor No. 2, and on April 4, TEPCO intentionally released radioactive water with lower contamination into the ocean, in order to make room to store more contaminated water.

With these two incidents only, the amount of radioactive release outside of the nuclear plant is 4,700 tera-becquerels (according to TEPCO estimate), 20,000 times the allowable limit.

This TEPCO estimate has been criticized by experts in and outside of Japan as "underestimation."

On December 4, water contaminated with 26 billion becquerels of radioactive strontium leaked into the ocean from the equipment that vaporizes and condenses processed contaminated water. It is estimated that the tanks to store processed contaminated water within the plant will be full by early next year. Water stored in those tanks contain radioactive strontium too. TEPCO has been considering further cleansing the water and releasing into the ocean. With opposition by fishery organizations, they have decided to postpone it.

NISA, in the interview with *Tokyo Shimbun*, emphasized the fact that their priority has been dealing with the accident, and the damage at the Fukushima Daiichi plant was a "state of emergency," one in which radioactive leaks could not be prevented. That was their reason for not applying the maximum allowable total amount of radioactive release and regarding the leak of 4,7000 tera-becquerels of radioactive material as "zero."

As NISA has stated this special treatment due to the "state of emergency" may continue until "the accident is brought under control," and they are ambiguous on when that exactly will be, saying it is a "matter to be discussed."

NISA has said that even when TEPCO releases radioactive water into the water, they will continue to regard the release as "zero."[49]

If this is the path to secure sea and human life, all of us are standing on a shimmering fiction: no disaster ahead.

PROTEST GOES INTERNATIONAL

With 11,500 participants from 30 countries attending the January 21–22, 2012, two-day Global Conference for a Nuclear Power Free World, delegates demanded a nuclear-free world and justice for Fukushima citizens after the thermonuclear explosions at three nuclear power plants in March 2011. A commonality shared by all was a vision of a world vastly different to the one that allowed the Fukushima nuclear disaster to occur—as well as a staunch determination to ensure that nuclear destruction never befalls our planet again.

Organized by several major Japan-based and international NGOs (with the Peace Boat serving as secretariat), the conference opening session began with one panelist after the next expressing dismay at Japan's present nuclear policies, as well as the suffering unleashed from the Fukushima Daiichi nuclear plant accident.

"Japan's nuclear regulatory agency being housed within the economic ministry is no different than criminals living under the same roof as the police," charged former Fukushima prefectural governor Eisaku Sato, long an outspoken critic of nuclear power.

Tetsunari Iida, the director of the Institute for Sustainable Energy Policies, spoke similarly of a "citizen revolution that has occurred as a result of the ludicrous, baseless statements that have continued to be released by the Japanese government following the Fukushima catastrophe"—a point echoed by German parliamentarian Rebecca Harms, who said pointedly, "With the majority public opinion in Japan now standing solidly against nuclear power, why the hell would Japan ever consider promoting it again?"

A documentary film project, *Uncanny Terrain*, follows organic farmers from Fukushima who have decided against all odds to continue their livelihoods—remaining on the land that has sometimes been in their families for generations, and continuing to produce food while testing it for safety standards.[50]

Consumers Protect Themselves

Reacting to radiation-tainted food, agricultural officials and many farmers fear that revealing more detailed results would scare away consumers, who might be spooked by even low levels of radiation. "We hear the calls for more disclosure, but revealing more detailed data would just hurt too many farmers," said Osamu Yoshioka, a food safety official at the Ministry of Agriculture. But so far, farmers say they have no choice but to try to sell their tainted produce because they have seen little if any compensation and must make a living. Fukushima Daiichi's operator, Tokyo Electric Power, has offered full compensation only to farmers in the zones that were evacuated, which were within twelve miles of the plant, and a larger area to the northwest, where winds carried much of the fallout.

This limitation with government lies have also had a corrosive effect on public confidence in the food-monitoring efforts, with a growing segment of the public and even many experts coming to believe that officials have understated or even covered up the true extent of the public health risk in order to limit both the recognized economic damage and the size of potential compensation payments.

Critics say farm and health officials have been too quick to allow food to go to market without adequate testing, or have ignored calls from consumers to fully disclose test results. And they say the government can no longer pull the wool over the public's eyes, as they contend it has done routinely in the past.

This attitude and limit on compensation gave birth to the Network of Parents to Protect Children from Radiation, known here as Mamorukai, which was started by a few dozen concerned parents after the accident. In nine months, it has grown into a nationwide network with two hundred chapters. "If the government treated us like adults, there would be no need for Mamorukai," said Sachiko Sato, a network founder. "Japan must build an entirely new food-monitoring system that we average people can really trust."[51]

Plant and animal life, too, must be respected, protected. In February 2012, the Ministry of the Environment started a study to examine the effects of radiation from the disaster-struck Fukushima No. 1 Nuclear Power Plant on nearby plants and wildlife, particularly their reproductive functions.

Working with the Japan Wildlife Research Center and the National Institute of Radiological Sciences, the Ministry began selecting local species of plants and animals based on twelve recommended for study by the International Commission on Radiological Protection covering in the wake of the nuclear disaster, including a type of mouse, shellfish, and pine tree. For comparison purposes, specimens were to be collected from areas with differing radiation levels both inside and near the no-entry zone around the plant. The specimens will be checked for radioactive cesium levels, and also for anomalies in their appearance or the impact on their DNA. The study will pay particular attention to the reproduction of amphibians and mammals and check for abnormalities in their offspring, as well as in the seeds of local flora.

If experience after the 1986 Chernobyl nuclear accident offers any indication of what might follow, abnormalities in the local swallow population were reported, such as low white blood cell counts, shrinking brain capacities, and a drop in bird numbers. Even twenty-six years later, radiation is said to be affecting the area. Without humans, local wildlife has been able to breed without restriction, but the effects of radiation on the ecosystem as a whole are still unknown.[52]

THE BIG FIX

Absolutely no one knows how to clean up the irradiated areas. But that has not deterred the Japanese government from starting to hand out an initial $13 billion in contracts meant to rehabilitate the more than 8,000-square-mile region most exposed to radioactive fallout—an area nearly as big as New Jersey. The main goal is to eventually enable the return of many of the 80,000 or more displaced people nearest the site of March's nuclear disaster, including the 6,500 villagers of little Iitate.

Consortiums led by Taisei and the other two big companies—Obayashi and Kajima among them—received contracts for the government's first twelve pilot decontamination projects, totaling about $93 million.

"It's a scam," said Kiyoshi Sakurai, a critic of the nuclear industry and a former researcher at a forerunner to the Japan Atomic Energy Agency, which is overseeing this phase of the decontamination effort. "Decontamination is becoming big business." The cleanup contracts, Mr. Sakurai and other critics contend, are emblematic of the too-cozy ties they say have long existed between the nuclear industry and government. "The Japanese nuclear industry is run so that the more you fail, the more money you receive," Mr. Sakurai said.

The Japan Atomic Energy Agency said the construction giants would not necessarily receive the bulk of the future work, which will be contracted out by the Environment Ministry. Officials for the companies, however, have indicated they expected to continue serving as primary contractors, as they had before, profiting hugely from Japan's previous embrace of nuclear power.

It was these same three companies that helped build forty-five of Japan's fifty-four nuclear plants—including the reactor buildings and other plants at Fukushima Daiichi that could not withstand the tsunami that caused the plant's catastrophic failure.[53]

National ministries meanwhile are overseeing most of the tsunami-hit area's large reconstruction projects out of a recently approved $120 billion budget. But Tokyo is handing $25 billion directly to regional and local governments to refashion their communities, a boon to politically connected construction companies. The thinking is that local officials understand their communities best, but local politicians and bureaucrats are also less likely to make tough decisions like sacrificing some villages to make others stronger—and to lower the reconstruction costs that are likely to sap already strained financial resources. Some critics, meanwhile, have argued that local companies and governments could perform the cleanup work for much less money, while creating local jobs.

BIG COMPANIES SACK THE NATIONAL BUDGET

The big construction meanwhile raided the cleanup allotments from the national budget. Big companies won the main contracts, the actual cleanup, tearing down structures, raking, digging, and scrubbing allotted to numerous subcontractors and sub-subcontractors, in turn relying on untrained day laborers to do the dirtiest decontamination work. This tiered system of extraction of revenues allows Japan's largest nuclear and construction industries to extract the lion's share of budgetary allotments, while others in the hierarchical system on down take ever lesser sums, as fees are siphoned off and wages dwindle each tier down the scale.

The pilot projects led by Taisei and other giant contractors immediately hit resistance, as the government failed to anticipate communities' reluctance to store tons of soil scraped from contaminated yards and fields. Besides, cleanup methods remained in doubt, as radioactive particles were swept by wind and rain, able to again contaminate towns after an initial cleanup. "No experts yet exist in decontamination, and there is no reason why the state should pay big money to big construction companies," said Yoichi Tao, a visiting professor in physics at Kogakuin University.

At the local level, this has brought on a generational battle. The reconstruction cities and fishing villages along the coast where the tsunami hit is a preview of what may be the most critical test Japan will face in the decades ahead. In a country where civic power rests disproportionately among older people, how does Japan, which has the world's most rapidly aging population, use its dwindling resources to build a society that looks to the future as much as to the past? The evolving debate over the shape of towns' reconstruction underscores how older Japanese, more attached to their land and customs, are wielding disproportionate influence and swaying local governments into issuing reconstruction blueprints at odds with Tokyo's stated goal of creating long-term sustainable communities.

The younger population has few choices—either to leave or, respecting their elders' decisions, quietly accept rebuilding to a past with little future opportunities for marriage partners or urban life.[54]

If forecasters hold a crystal ball, their future offers little. For Japan has been the world's fastest aging country, and with its birthrate among the lowest, its population decline facing the deepest global slide in coming decades. Experts say that Japan's population will keep losing 1 million every year in coming decades and the country urgently needs to overhaul its social security and tax system to reflect the demographic shift to pay to support its aging population. Japan's rapid aging means the national population of 128 million will shrink by one-third by 2060, and seniors will account for 40 percent of people, placing a greater burden on the shrinking population of workers to support the social security and tax systems.

The population estimate by the Health and Welfare Ministry paints a grim future: In year 2060, Japan will have 87 million people. The number of people 65 or older will nearly double to that Golden Age of 40 percent, while the national workforce of people between ages 15 and 65 will shrink to about half of the total population, according to the estimate, made by the National Institute of Population and Social Security Research.

The total fertility rate, or the expected number of children born per woman during her lifetime, in 2060 is estimated at 1.35, down from 1.39 in 2010—well below more than 2 needed to keep the country's population from declining. But the average Japanese will continue to live longer. The average life expectancy for 2060 is projected at 90.93 for women, up from 86.39 in 2010, and 84.19 years for men, up from 79.64 years.[55]

PROTECT OUR CHILDREN!

There are more dangerous consequences, too.

A new study from the Radiation and Public Health Project found that babies born in the western United States as well as other Pacific countries shortly after the Fukushima nuclear disaster in Japan in March 2011 may be at greater risk for congenital hypothyroidism.

Babies born in places including Hawaii, Alaska, California, Oregon, and Washington shortly after Fukushima were 28 percent more likely to suffer from the illness, according to the study, than children born in those same regions one year earlier. The illness, if untreated, can cause permanent handicaps in both the body and brain.

According to the U.S. National Library of Medicine, "If untreated, congenital hypothyroidism can lead to intellectual disability and abnormal growth. In the United States and many other countries, all newborns are tested for congenital hypothyroidism. If treatment begins in the first month after birth, infants usually develop normally."

MSN's Healthy Living blog explains the Fukushima explosions led to clouds of radioisotope iodine-131 that "floated east over the Pacific Ocean and landed through precipitation on West Coast states as well as other Pacific countries."

In Japan, the health effects associated with Fukushima are obviously much worse. The mortality rate of elderly people who were in retirement facilities near the nuclear plant has reportedly tripled. There has also been reported increases in the number of children with flat feet, thought to be the result of kids playing on radiated soil.

Experts suggest that parents of children born in the western United States or Pacific regions in March or April 2011 get their children checked by a pediatrician for congenital hypothyroidism.[56]

NO END IN SIGHT

The present as history is hard to grasp.

Almost immediately after World War II, the U.S. and France imposed the equipment and fuels for nuclear production of electrical energy on Japan, accomplished through a CIA alliance with conservative wartime politicians and business operatives.

U.S. corporations like General Electric made the plants, made the profits and tried to keep Japan on a technological leash for using American-provided plutonium fuel rods, plant design, and location of the plants near the sea to provide coolant in case of disasters from the tremendous heat the plants emit. But Japanese nuclear utilities broke free of U.S. controls, building an unregulated apparatus of politicians, business elites, and compliant workers living on the edge of earthquakes, tsunamis, and the destruction of nuclear plants, risking the lives of millions of unsuspecting people—ordinary people, who held the vaguest information about the time bombs in their midst.

The nuclear explosions at Daiichi proved the worst disaster in modern Japanese history, second only to the U.S. atomic bombs rained on Hiroshima and Nagasaki. Now the Japanese people are being forced to pay the awful price, while the power companies, construction corporations, and builders of future nuclear plants reap the rewards.

The grim reaper will take a continuous future toll in lives and health from the highly radioactive wastewater surging from Fukushima Daiichi even now.

As the dangers emerge and reemerge, know there is no future anywhere near the fallen plant. For every minute almost 285 liters, or 75 gallons, of groundwater is percolating into the plant's ravaged reactor buildings, where it becomes contaminated before being pumped out from swamping a critical cooling system.

To rely on 940 hulking steel and leaking storage tanks, spread across 17 hectares, some 42 acres, of parking lots and lawns remain insufficient to handle the many tons of strontium-laced water.

In May 2013, TEPCO planned still another disaster—to cut down an entire forest on the southern edge of the Daiichi plant to create room for hundreds of new tanks to store the water overflow leaking from underground pits, threatening to escape into the Pacific Ocean.[57]

And when TEPCO discovered 0.39 becquerels of radioactive cesium-137 per liter of water leaking from a single storage tank—an amount that is far below Japan's safety level for drinking water of 10 becquerels per liter—fishermen said they believed absolutely nothing of TEPCO newspeak about dumping nonirradiated water into the sea.[58]

Ocean simulations showed that the plume of radioactive cesium-137 released by the Fukushima disaster in 2011 could begin flowing into U.S. coastal waters starting in early 2014 and peak in 2016. Luckily, two ocean currents off the eastern coast of Japan—the Kuroshio Current and the Kuroshio Extension—would have diluted the radioactive material so that its concentration fell well below the World Health Organization's safety levels within four months of the Fukushima incident. But it could have been a different story if nuclear disaster struck on the other side of Japan. "The environmental impact could have been worse if the contaminated water would have been released in another oceanic environment in which the circulation was less energetic and turbulent," said Vincent Rossi, an oceanographer and postdoctoral research fellow at the Institute for Cross-Disciplinary Physics and Complex Systems in Spain.[59]

A NEVER-ENDING BATTLE

The conditions at Daiichi remain a never-ending battle of imperious words and lies.

Under public pressure and disbelief that he is glossing over the truth, Pime Minister Shinzo Abe stood by his reassurances about Tokyo's safety that he gave to the International Olympic Committee in September before the city was awarded the right to host the 2020 Summer Olympic Games.

To appease massive public criticism, on September 26th he ordered the TEPCO operator of Japan's crippled nuclear power plant to scrap all six reactors at the Daiichi site instead of just four scheduled for decommissioning and to concentrate on tackling pressing issues like leaks of radioactive water.

Officials had long acknowledged that radiation-contaminated groundwater had been seeping into the ocean since soon after meltdowns and explosions crippled the plant. But Abe nonetheless insisted that radioactive water had been contained at the Daiichi plant and said that he would fend off "rumors" regarding Fukushima's safety.[60]

It was a complete lie. No thinking Japanese could possibly believe him.

ANOTHER TERRIBLE MOMENT

Another terrible moment was also soon to pass.

More than 6,000 fuel rod assemblies slumber in a common water pool just 50 meters from Daiichi's Unit 4. The pool, over 100 feet off the ground with no containment cover over it, is seeded with plutonium and other nuclides—extremely vulnerable to fission and explosion if it loses coolant.

According to Arnie Gundersen (the nuclear engineer with forty years' experience in an industry manufacturing fuel rods) the spent rods in the Unit 4 core are bent, damaged and brittle to the point of crumbling. Attempting to remove them might expose them to air, catch fire, and release unknown charges of radiation into the atmosphere.

Worst still, according to nuclear expert and former U.S. Department of Energy official Robert Alvarez, more than 11,000 fuel assemblies are scattered around the Daiichi site, with more than 85 times as much lethal cesium as was released at Chernobyl.

And what if the pool comes crashing to the ground, dumping the rods together into a pile as happened at Chernobyl, causing immediate fission.

The resulting radioactive cloud would threaten the health and safety of the people not only in Fukushima and throughout Japan, but also the entire planet. *The world community must now take charge at Fukushima.*[61]

Take it as you must. There is no exit now or in the future from the Japanese government and corporate lies about the Nuclear Tsunami and its spread.

There also is no place to hide from Daiichi's irradiated soils, tainted foods, crumbling structures, and wafting plutonium winds—possibly for the coming 25,000 years! Medical science can try to protect those injured, especially children. But this lethal, historical moment is set in place; and disability, pain, and early death await millions of innocents.

If we can grasp this, we know what we must do in the future to save our children and our species, among others that share our only planetary home.

NOTES

1. Martin Fackler, "Japanese Split on Fukushima Radiation Cleanup," *New York Times*, December 7, 2011.
2. Richard Krooth and Hiroshi Fukurai, *Common Destiny: Japan and the United States in the Global Age* (Jefferson, N.C: McFarland & Co., 1990).
3. Martin Fackler, "Japan Rethinks a Dam, and a Town Protests," *New York Times*, October 15, 2009.
4. "Disaster in Japan," *the Economist*, April 21, 2011.
5. Fackler, supra note 3.
6. "A Crisis of Leadership, Too," *The Economist*, April 26, 2011.
7. "Tepco Confirms Extra Partial Fuel Rod Meltdown at Plant," *News Business* (BBC, May 23, 2011).
8. Jan Beránek, "Japan to Abandon Nuclear Plans and Embrace Renewable Energy—Who's Next?," May 12, 2011.
9. David Wagner, "Report From Tokyo: Who to Believe?," *The Blog, Huffington Post*, May 27, 2011, http://www.huffingtonpost.com/david-wagner/report-from-tokyo-who-to-_b_867863.html.
10. James West, "Uranium Stocks Under Siege as Fukushima Continues Spewing Radiation," May 30, 2011, MidasLetter.com.
11. "Canada's Port Hope Reaps Uranium's Rewards and Risks," Reuters/News, June 2, 2011, News@Reuters.com.
12. "Broadcast Survey" (Fuji Television Network, May 28, 2011).
13. Bettina Wassener, "Moody's Warns That It Might Downgrade Japan's Debt," *New York Times*, May 31, 2011, sec. The Global Business.
14. Hiroko Tabuchi, "Mounting Woes Prompt Sony to Predict Another Annual Loss," *New York Times*, November 2, 2011.
15. Hiroko Tabuchi, "Japan Acts Alone to Weaken Its Currency," *New York Times*, October 31, 2011.
16. Hiroko Tabuchi, "Strong Yen Is a Two-Edged Sword for Japan," *New York Times*, August 19, 2011.
17. Rob Cox, "Japan's Misguided Foreign Acquisitions, REUTERS BREAKING VIEWS," *New York Times*, June 2, 2011, http://www.nytimes.com/2011/06/03/business/03views.html?emc=tnt&tntemail1=y'.
18. Martin Fackler, "Public Finance Minister Is Chosen as Japan's Next Leader," *New York Times*, August 29, 2011.

19. Hiroko Tabuchi, "World/Asia Pacific: Japan Joining Talks on Trans-Pacific Partnership, Noda Says," *New York Times*, November 12, 2011.

20. Martin Fackler, "Radiation Discovered in Rice Near Tokyo," *New York Times*, August 19, 2011.

21. Naoko Fujimura and Chris Cooper, "Issue of Radiation-Tainted Food in Japan Escalates, Mushrooms Join Growing List of Radioactive Threats to Japan's Food Chain," *Vancouver Sun*, August 14, 2011.

22. Hiroko Tabuchi, "Japanese Tests Find Radiation in Infant Food," *New York Times*, December 6, 2011.

23. Naoko Fujimura and Chris Cooper, "Issue of Radiation-Tainted Food in Japan Escalates, Mushrooms Join Growing List of Radioactive Threats to Japan's Food Chain," *Vancouver Sun*, August 14, 2011. Ibid.

24. Hiroko Tabuchi, "Japan Lifts Ban on Beef from Disaster Area," *New York Times*, August 25, 2011.

25. http://ajw.asahi.com/article/behind_news/social_affairs/AJ201111240030

26. Tomohiro Iwata, "TEPCO: Radioactive Substances Belong to Landowners, Not Us," *Asahi Shimbun Weekly AERA*, n.d., http://ajw.asahi.com/article/behind_news/social_affairs/AJ201111240030.

27. Miki Tanikawa, "Finding Use for 25 Million Tons of Rubble," *New York Times*, June 2, 2011, Global Business edition, http://www.nytimes.com/2011/06/03/business/global/03iht-RBOG-JAPAN-RUBBLE03.html?emc=tnt&tntemail1=y].

28. "Can It Happen Here? Yes—if We Let It. Fukushima's Lessons for the United States," *Catalyst*, Summer 2011, http://www.ucsusa.org/assets/documents/catalyst/catalyst-summer-2011.pdf.

29. Martin Fackler, "After Kan, Instability Looms as Japan's Greatest Threat," *New York Times*, August 26, 2011.

30. Martin Fackler, "A Governor's Power To Shape the Future Of a Nuclear Japan," *New York Times*, July 3, 2011.

31. Tomooki Yasuda, "A Report Suggests a Second Meltdown at Reactor at Fukushima Plant," *Ashia Shimbus*, August 22, 2011, http://ajw.asahi.com/article/0311disaster/fukushima/AJ201108085674.

32. Hiroko Tabuchi, "In Japan, a Tenuous Vow to Cut," *New York Times*, September 1, 2011.

33. Hiroko Tabuchi, "Japanese Parliament Backs Noda as Prime Minister," *New York Times*, August 30, 2011.

34. http://www.beyondnuclear.org/

35. "Environmental Radiation: Where the Radiation Went," *Nature* 477, no. 7362, 9 (September 1, 2011), http://dx.doi.org/10.1038/477009a.

36. Hiroko Tabuchi, "Japan's Premier Says He'll Keep Nuclear Phase-Out," *New York Times*, September 3, 2011.

37. Norimitsu Onishi and Martin Fackler, "Utility Reform Eluding Japan After Nuclear Plant Disaster," *New York Times*, November 8, 2011.

38. "Japan's Disaster: A Crisis of Leadership, Too," April 26, 2011, http://www.economist.com/node/18441143.

39. Hiroko Tabuchi, "Japanese Island's Activists Resist Nuclear Industry's Allure," *New York Times*, August 28, 2011.

40. Martin Fackler, "World/Asia Pacific: Japan Set to Declare Control Over Damaged Nuclear Reactors," *New York Times*, December 15, 2011.

41. "Typhoon Rains Kill at Least 25 and Maroon Thousands in Japan," Associated Press, September 4, 2011.

42. Hiroko Tabuchi and Martin Fackler, "More Radioactive Water Leaks at Japanese Plant," *New York Times*, December 4, 2011.

43. Martin Fackler, "Japan Says Decommissioning Damaged Reactors Could Take 40 Years," *New York Times*, December 21, 2011.

44. Martin Fackler and Ko Sasaki, "Japan Official Ordered Nuclear E-Mails, Inquiry Finds," *New York Times*, September 8, 2011.

45. So instead of being caused by self-sustaining nuclear fission, the xenon found in the Number 2 reactor, TEPCO said it was likely generated by spontaneous fission of curium and other radioactive substances derived from the reactor fuel,. Spontaneous fission allegedly takes place in suspended nuclear reactors under normal conditions. "Experts back TEPCO's view that criticality did not occur,' *Asahi Shimbun*. November 3, 2011] And a naturally occurring element, boron, supposedly soaks up the neutrons released when an atom is split so that those neutrons cannot go on to split other atoms when material "goes critical" in the process of fission. See also Hiroko Tabuchi, "Fears of Fission Rise at Stricken Japanese Plant," *New York Times*, November 2, 2011.

46. Hiroko Tabuchi, "World/Asia Pacific: Fukushima Meltdown in Japan May Have Been Worse Than Thought," *New York Times*, November 30, 2011.

47. "Alec Baldwin Knocks Nuclear Power, Calls Reactors 'Filthy,'" *Huffington Post/HufPost Green*, September 4, 2011, http://www.huffingtonpost.com/2011/09/02/alec-baldwin-nuclear-power-dirty-energy_n_946555.html.

48. Tons of debris from Japan will likely begin washing ashore in about a year, from California to southern Alaska, they said. Items that wash up may include portions of houses, boats, ships, furniture, portions of cars, and just about anything else that floats; See AP, "Flotsam from Japanese tsunami reaches West Coast," December 15, 2011, http://www.sfgate.com/cgi-bin/article.cgi?file=/n/a/2011/12/15/national/a105013S15.DTL

49. Japan's Nuclear Safety Agency Regards Radioactive Water Leaks to the Ocean as "Zero," Shimbun Report (Tokyo, Japan, December 1, 2011).

50. http://tenthousandthingsfromkyoto.blogspot.com/2012/01/passion-runs-high-at-global-yokohama.html

51. Martin Fackler, "World / Asia Pacific: Wary, Japanese Take Food Safety Into Their Own Hands," *New York Times*, January 22, 2012.

52. "Gov't Starts Study of Radiation Effects on Wildlife Near Fukushima Plant," *Mainichi Daily News*, February 12, 2012, http://mdn.mainichi.jp/mdnnews/news/20120130p2a00m0na011000c.html.

53. Hiroko Tabuchi, "Japan Starts Nuclear Cleanup, With Little Idea of How," *New York Times*, February 10, 2012.

54. Norimitsu Onishi, "As Japan Works to Patch Itself Up, a Rift Between Generations Opens," *New York Times*, February 12, 2012, http://www.nytimes.com/2012/02/13/world/asia/amid-japan-reconstruction-generational-rift-opens.html?_r=1&emc=tnt&tntemail1=y.

55. Mari Yamaguchi, "Japan Population to Shrink by One-third by 2060," *MSN News/Associated Press*, January 30, 2012, http://news.ca.msn.com/world/japan-population-to-shrink-by-one-third-by-2060.

56. Mike Krumboltz, "Fukushima Fallout May Be Causing Illness in American Babies: Study," Yahoo News, April 5, 2013.

57. Martin Fackler, "Fukushima Plagued by Radioactive Wastewater." *International Herald Tribune*, April 30, 2013.

58. Martin Fackler, "Leak Found in Steel Tank for Water at Fukushima," *New York Times*, June 5, 2013.

59. Jeremy Hsu, "Fukushima's Radioactive Plume Could Reach U.S. Waters By 2014," *LiveScience*, September 1, 2013.

60. "Japan: Fukushima's Surviving Reactors Ordered Shut," AP, September 19, 2013.

61. Harvey Wasserman, "Humankind's Most Dangerous Moment: Fukushima Fuel Pool at Unit 4: "This is an Issue of Human Survival," *Global Research*, September 20, 2013.

Bibliography

"20 Millisieverts for Children and Kosako Toshiso's Resignation," *Asian Pacific Journal: Japan Focus*, May 01, 2011.

"20 Seiki Media Kenkyujo Tokubetu Kenkyukai: CIA to Ogata Taketora [20 Century Media Research Institute, Special Study Session: CIA & Taketora Ogata]" July 25, 2009, http://members.jcom.home.ne.jp/katote/0907OGATA.pdf.

"A Crisis of Leadership, Too," *The Economist*, April 26, 2011.

"Activation of Echelon Units," in *History of the Air Intelligence Agency, January 1-December 31, 1994* (San Antonio, TX: AIA, 1995).

"Alec Baldwin Knocks Nuclear Power, Calls Reactors 'Filthy'," *Huffington Post*, September 4, 2011.

Aoki, Mizuho. "Thyroid Cancers Up in Fukushima," *Japan Times*, Nov. 11, 2013.

Arima, Tetsuo. Genpatsu, Shoriki, CIA: Kimitsu Bunsho de Yomu Showashi [Nuclear Power, Shoriki, & CIA: History of Showa from Secret Documents] (Tokyo: Shincho-Shinsho, 2008).

Arima, Tetsuo, "Reisen no Media Nippon Terebi Hosomo: Shoriki, Maikurowe-bumo o Meguru Geikoku Hankyosanshugi Kaiko Johoseisaku [Media in Cold War and Nippon TV Network: Shoriki & America's Anti-Communism Foreign and Intelligence Policies," www.f.waseda.jp/tarima/coldwarNTV.pdf.

Arima, Tetsuo. *Showashi o Ugokashita Amerika Johokikan (American Intelligence Agencies that Influence the History of Showa)* (Tokyo: Shincho-Shinsho, 2009).

Arima, Tetsuya. "Cold War, East Asia, and Television," *Cultural Studies Monthly*, August 25, 2007.

"Atomic Cover-Up: The Hidden Story Behind the U.S. Bombing of Hiroshima and Nagasaki," *Democracy Now*, Aug. 9, 2011.

Baker, Scott, "'Hail Mary Pass' Japan's Frantic Efforts to Avoid Nuclear Meltdown," *Blaze*, Mar. 12, 2011.

BEIR (Biological Effects of Ionizing Radiation), *Health Effects of Exposure to Low Levels of Ionizing Radiation: BEIR V* (Washington, D.C.: National Academy Press, 1990).

Belson, Ken and Steve Lee Myers, "Tokyo Utility Lays Out Plan for Its Reactors," *New York Times*, April 17, 2011.

Beránek, Jan. "Japan to Abandon Nuclear Plans and Embrace Renewable Energy—Who's Next?" *Bruno Santos*, May 12, 2011.

Beránek, Jan, Yukiya Amano, and Phillip White, "U.N. Nuclear Watchdog Says It Will Continue to Push for New Nuclear Power Plants Despite Growing Global Nuclear Concern," *Democracy* Now, April 5, 2011.

Bergé, Christine. "Nuclear Debate: Superphénix Embers Under the Ashes," *Le Monde Diplomatique*, April 2011, http://www.monde-diplomatique.fr/2011/04/BERGE/20399
Berger, John J. *Nuclear Power: The Unviable Option* (Palo Alto, California: Ramparts Press, 1976).
Black, Richard. "Radiation Falls at Japanese Plant," BBC, March 15, 2011.
Bossew, Peter, et al, *35 Years Promotion of Nuclear Energy: International Atomic Energy Agency; A Critical Documentation of the Agency's Policy*. (Vienna-Hannover: Gruppekologie,: AntiAtom International and Kologie Institut., 1993).
Boyer, Dave. "NSA Spying Sours Relations with Germany, Brazil," *Washington Times*, Oct. 24, 2013.
Bradsher, Keith. "Japan Considers Tightening Access Around Nuclear Plant," *New York Times*, April 20, 2011.
Brain, Marshall. "Blast from the Past: Our Friend the Atom," *Howstuffworks*, April 15, 2010.
Brasor, Philip. "Nuclear Policy was Once Sold by Japan's Media," *Japan Times*, May 22, 2011.
Breidthardt, Annika. "German Government Wants Nuclear Exit by 2022 At Least," Reuters, May 30, 2011.
"Broadcast Survey." Fuji Television Network, May 28, 2011.
Brumfiel, Geoffrey. "World Health Organization Weighs in on Fukushima," *Newsblog*, May 23, 2013.
Buerk, Roland. "Japan's Government Approves Tepco Compensation Scheme," *BBC*, May 13, 2011.
Burke, Matthew M. "Congress Wants Answers on Health Impacts of Japan Disaster Relief," *Stars & Stripes*, Jan. 27, 2014.
Caldicott, Helen. *Nuclear Madness: What You Can Do* (New York and London: W. W. Norton and Company, 1994).
"Canada's Port Hope Reaps Uranium's Rewards and Risks," Reuters/News, June 2, 2011.
Caroll, Chris. "Pentagon Preparing for a Nuclear Worst-Case Scenario at Fukushima," *Stars & Stripes*, Mar. 16, 2011.
Carroll, Sean B. *Endless Forms Most Beautiful: The New Science of Evo Devo* (New York: W.W. Norton, 1995).
Central Intelligence Agency, "A Look Back . . . Robert Carey Broughton: From Walt Disney to War Movies," September 4, 2009, https://www.cia.gov/news-information/featured-story-archive/robert-carey-broughton.html.
Central Intelligence Agency, *Intelligence in New Japan*, September 18, 1995.
CIA to Sengo Nihon [CIA & Post-war Japan] (Tokyo: Shincho-Shinsho, 2010).
Committee to Assess Health Risks from Exposure to Low Levels of Ionizing Radiation, National Research Council, Health Risks from Exposure to Low Levels of Ionizing Radiation: BEIR VII Phase 2 (Washington, DC: National Academies Press, 2006).
"Container Damaged, Radiation Leak Feared at Fukushima No. 2 Reactor," *Kyodo News*, Mar. 15, 2011.
Cousteau, Jacques. "Periodic Essays: The Bounty of the Sea," *Deviant Art*, 1994, http://lemon-sauce.deviantart.com/journal/The-Bounty-of-the-Sea-by-Jacques-Cousteau-231816293.
Cox, Rob. "Japan's Misguided Foreign Acquisitions, REUTERS BREAKING VIEWS," *New York Times*, June 2, 2011.
Craig, William. *The Fall of Japan* (New York: Dell Publishing Co., 1967).
Dalton, Russell, et al., *Critical Masses: Citizens, Nuclear Weapons Production, and Environmental Destruction in the United States and Russia* (Massachusetts: MIT Press, 1999).
"Damage to the Unit 4 Nuclear Reactor Building at Fukushima Daiichi Nuclear Power Station," *TEPCO*, Mar. 15, 2011.
Dedman, Bill. "General Eectric-Designed Reactors in Fukushima Have 23 Sisters in U.S.," *NBC News*, Mar. 12, 2011.
"'Depleted' Uranium Munitions: Nuclear Waste as a Weapon," *Military Toxics Project: Information Sheet*, June 2003.
"Disaster in Japan," *The Economist*, April 21, 2011.
Dusinberre, Martin "How Japan's Nuclear Industry Got Here," Reuters, March 18, 2011.

Dutta, Ratnajyoti. "India Bans Japan Food Imports, Says Radiation Spreading," Reuters/New Delhi, April 5, 2011.

Edano, Yukio, "Japan to Release 11,000 Tons of Radioactive Water into Ocean," *Democracy Now*, April 4, 2011.

Eichelberg, Don. "Public Is Opposed, Industry Pushes On," *Nuclear Guardianship Forum*, no. 3, Spring 1994.

Eisler, Peter. "Contaminated Water Heightens Concerns Over Nuclear Plant," *USA Today*, March 25, 2011.

"Ex-Bureaucrat Details Secret U.S. Nuke Pact," *Japan Times*, June 30, 2009.

Fackler, Martin. "A Governor's Power To Shape the Future Of a Nuclear Japan," *New York Times*, July 3, 2011.

Fackler, Martin. "After Kan, Instability Looms as Japan's Greatest Threat," *New York Times*, August 26, 2011.

Fackler, Martin. "Fukushima Plagued by Radioactive Wastewater." *International Herald Tribune*, April 30, 2013.

Fackler, Martin. "Japan Reaffirms Nuclear Energy Use," *New York Times*, May 8, 2011.

Fackler, Martin. "Japan Rethinks a Dam, and a Town Protests," *New York Times*, October 15, 2009.

Fackler, Martin. "Japan Says Decommissioning Damaged Reactors Could Take 40 Years," *New York Times*, December 21, 2011.

Fackler, Martin. "Japan's Prime Minister Defends Handling of Crisis," *New York Times*, Apr. 30, 2011.

Fackler, Martin. "Japanese Split on Fukushima Radiation Cleanup," *New York Times*, December 7, 2011.

Fackler, Martin. "Leak Found in Steel Tank for Water at Fukushima," *New York Times*, June 5, 2013.

Fackler, Martin. "Otsuchi. Before It Can Rebuild, Japanese Town Must Survive," *New York Times*, May 15, 2011.

Fackler, Martin. "Public Finance Minister Is Chosen as Japan's Next Leader," *New York Times*, August 29, 2011.

Fackler, Martin. "Radiation Discovered in Rice Near Tokyo," *New York Times*, August 19, 2011.

Fackler, Martin. "World/Asia Pacific: Japan Set to Declare Control Over Damaged Nuclear Reactors," *New York Times*, December 15, 2011.

Fackler, Martin. "World/Asia Pacific: Wary, Japanese Take Food Safety Into Their Own Hands," *New York Times*, January 22, 2012.

Fackler, Martin and Ko Sasaki, "Japan Official Ordered Nuclear E-Mails, Inquiry Finds," *New York Times*, September 8, 2011.

Fitsanakis, Joseph. "Documents Reveal CIA Meddling in Japanese Elections," *Intelnews.org*, July 28, 2009.

Flac, T. D. Travis Tritten, and Erik Slavin, "First Military Evacuation Flight Leaves Japan," *Stars & Stripes*, Mar. 19, 2011.

"Former Prime Minister Nakasone Receives 'Atoms for Peace' Prize," *Atoms in Japan: News in Brief*, Oct. 29, 2008.

"Former US Army Depleted Uranium Expert Alleges Campaign of Harassment Against him for Speaking Out," *Democracy Now*, May 30, 2000.

Fujimura, Naoko and Chris Cooper. "Issue of Radiation-Tainted Food in Japan Escalates, Mushrooms Join Growing List of Radioactive Threats to Japan's Food Chain," *Vancouver Sun*, August 14, 2011.

"Fukushima-Area Cancer Risks Are Higher Than Normal After Japan Nuclear Disaster, WHO Reports," *Huffington Post*, Feb. 6, 2014.

"Fukushima Radiation Levels Will Concentrate in Pockets at Specific U.S. and Canada West Coast Locations," *Global Research News*, Nov. 2, 2013.

Glanz, James and William J. Broad, "U.S. Sees Array of New Threats at Japan's Nuclear Plant," *New York Times*, April 5, 2011.

Glionna, John M., "Anger and Abandonment in a Japanese Nuclear Ghost Town," *Los Angeles Times*, March 31, 2011.

Goffman, John W. and Egan O'Connor, "The Law of Concentrated Benefit Over Diffuse Injury" (presented at the Committee for Nuclear Responsibility, San Francisco:, 1994).

Gonzalez, Juan. "Army to Test N.Y. Guard Unit," *New York Daily News*, Apr. 5, 2004.

Gonzalez, Juan. "Report from Sendai: Fears of Radioactivity Are Hampering Relief Efforts," *Democracy Now*, March 17, 2011.

Goodman, Amy. "Nuclear Catastrophe in Japan "Not Equal to Chernobyl, But Way Worse," *Democracy Now*, April 12, 2011.

Gordon, Julie. "Canada's Port Hope Reaps Uranium's Rewards and Risks," *Reuters U.S.*, June 2, 2011.

"Gov't Starts Study of Radiation Effects on Wildlife Near Fukushima Plant," *Mainichi Daily News*, February 12, 2012.

Greenemeler, Larry. "How Does an Earthquake Trigger Tsunamis Thousands of Kilometers Away?" *Scientific American*, Mar. 11, 2011.

Grunden, Walter E., Mark Walker, and Masakatsu Yamazaki, "Wartime Nuclear Weapons Research in Germany and Japan," *OSIRIS*, 20:107-130, 2005.

Gucciardi, Anthony. "Is the Government Stockpiling Iodide in Preparation for Fukushima Meltdown?" *Storyleak*, Jan. 1, 2014.

Gundersen, Arnie. "This Could Become Chernobyl on Steroids," *Democracy Now*, March 15, 2011, http://www.democracynow.org/2011/3/15/this_could_become_chernobyl_on_steroids.

Hacker, Barton C. *Elements of Controversy: The Atomic Energy Commission and Radiation Safety in Nuclear Weapons Testing 1947–1974* (Berkeley: UC Press, 1984).

Harlan, Chico. "After Japan's Tsunami, a Town Climbs Back," *Washington Post*, May 10, 2011.

Harlan, Chico, and Brian Vastag, "Radiation Levels at Japan Nuclear Plant Reach New Highs," *Washington Post*, Mar. 27, 2011.

Harer, Stephen. "At a Two Year Anniversary, the Tohoku Earthquake/Tsunami and Fukushima Nuclear Disaster Continue to Exact a Toll," *Forbes*, March 11, 2013.

Hashikawa, Bunzo. "Kakushin Kanryo [Reformist Bureaucrats]," in Jiro Kamishima (ed.), *Kenryoku no Shiso* (Tokyo: Chikuma Shobo, 1965).

Hatch, M. C. et al. "Cancer Near the Three Mile Island Nuclear Plant: Radiation Emissions," *American Journal of Epidemiology* Vol. 132, no. 3 (1990).

Hayes, Peter. "Should the United States Supply Light Water Reactors to Pyongyang?" *Nautilus Pacific Research*, November 16, 1993.

Hedges, Chris. *Death of the Liberal Class* (NY: Nation Books, 2010).

Heyes, J. D. "Radiation Dunce Dr. Shinichi Yamashita Finally Admits Giving Bad Info About Fukushima Fallout," *Natural News*, Nov. 22, 2013.

Higgens, Andrew. "Amid Nuclear Crisis, Japan's Tepco Planned New Reactors," *Washington Post*, April 6, 2011.

"Hiroshima to Fukushima: The Political Background to the Nuclear Disaster in Japan," *World Socialist*, June 23, 2011.

Hoppin, Jason. "State Rebuffs Radiation Concerns at Beach," *Santa Cruz Sentinel*, Jan. 9, 2014.

Hosaka, Tomoko A. "Workers Enter Damaged Reactor Building for 1st Time Since Right After Earthquake," *Huffington Post*, May 5, 2011.

Hsu, Jeremy. "Fukushima's Radioactive Plume Could Reach U.S. Waters By 2014," LiveScience, September 1, 2013.

"Hundreds Sue Toshiba, GE, Hitachi Over Fukushima Nuclear Disaster," *The Voice of Russia*, Jan 31, 2014.

Ijichi, Ken and Moshe Komata, "Efforts to Plug Japanese Reactor Leak Seem to Fail," *New York Times*, April 3, 20011.

International Atomic Energy Agency, "A Short History of IAEA," http://www.iaea.org/About/history.html (last visited October 10, 2011).

International Physicians for the Prevention of Nuclear War / The Institute for Energy and Environmental Research, *Plutonium: Deadly Gold* (Cambridge: International Physicians Press, 1992).

Ito, Shingo. "Radiation Levels Had Soar at Japan Nuclear Plant," *Agence France-Presse*, March 11, 2011.

Iwata, Tomohiro. "TEPCO: Radioactive Substances Belong to Landowners, Not Us," *Asahi Shimbun Weekly AERA*, 2011.

"Japan: Fukushima's Surviving Reactors Ordered Shut," Associated Press, September 19, 2013.

"Japan's Disaster: A Crisis of Leadership, Too," *Economist*, April 26, 2011.

"Japan's Energy Pact with Turkey Raises Nuclear Weapons Concerns," *Asahi Shimbun*, Jan. 7, 2014.

"Japan's Nuclear Safety Agency Regards Radioactive Water Leaks to the Ocean as "Zero," *Shimbun Report*, December 1, 2011.

"Japan to Abandon Troubled Fast Breeder Reactor," *APF*, Feb. 7, 2014.

"Japan to Release Revised Roadmap to Stabilize Crippled Nuclear Plant, Help Evacuees," *Associated Press/The Canadian Press*, May 16, 2011.

Japanese Government. *Report of Japanese Government to the IAEA Ministerial Conference on Nuclear Safety*, June 2011.

Jinzaburo, Takagi. "Critique of Japan's Nuclear Energy Program: Collected Papers of Jinzaburo Takagi, Tokyo." Citizens' Nuclear Information Center, 1994.

Johnson, Chalmers. *Nemesis: The Last Days of the American Republic* (New York: Metropolitan Books, 2006).

Johnson, Eric. "Key Players Got Nuclear Ball Rolling," *Japan Times*, July 16, 2011.

Johnston, Eric. "Operation Tomodachi a Huge Success, But Was It a One-Off?" *Japan Times*, Mar. 3, 2012.

Jolly, David, Hiroko Tabuchi, and Keith Bradsher, "Tainted Water at 2 Reactors Increases Alarm for Japanese," *New York Times*, Mar. 27, 2011

"Kakuo Motometa Nihon [Japan that Sought Nuclear Weapon]," *NHK Sukupu Dokumento*, October 3, 2010.

"Kan Sticks to Nuke Crisis Road Map: Meltdown Won't Halt Plan to Tame Plant by January, Leader Says," *Japan Times*, May 17, 2011.

Kanmatsu, Ichizo *Nihon Terebi Hosomo Koso o Shoriki Matsutaro* [Nippon Television Network Projects and Matsutaro Shoriki] (Mie, Japan: Mie University Press, 2005).

Kaplan, David E., U.S. Propaganda Efforts in Postwar Japan, Japan Policy Research Institute, February, 1997.

Katz, Richard. "How Japan's Nakasone Should Play His Cards Against Henry Kissinger," 48 *Executive Intelligence Review*, 47, 49 (1982).

Kingston, Jeffrey. *Tsunami: Japan's Post-Fukushima Future* (New York: Foreign Policy Magazine, 2011).

Koide, Hiroaki. Genpatsu no Anzen Shinwa o Kangaeru [Critique of the Safety Myth of Nuclear Energy], September 4, 2011. https://www.youtube.com/watch?v=vtrSMl0Q0sM.

Koide, Hiroaki, et al., *Saiakuno Kakushisetsu Rokkasho Saishori Kojo* (Tokyo: Shueisha, 2012).

Komanoff, Charles, and Cora Roelofs, *Fiscal Fission: The Economic Failure of Nuclear Power* (Washington, DC: Greenpeace, 1992).

"Koriyama Shimin—Hoshanokenen no Gakkojido Hihan Uttae Ga Saibansho De Kyakka [Koreyama Citizens: Court Denial of Plea to Relocate Children Due to Radiation," *Voice of Russia*, Apr. 25, 2013.

Kragh, Helge. *Quantum Generations: A History of Physics in the Twentieth Century* (Princeton: Princeton University Press, 1999).

Krooth, Richard, and Hiroshi Fukurai. *Common Destiny: Japan and the United States in the Global Age*. Jefferson, N.C.: McFarland & Co., 1990.

Krumboltz, Mike. "Fukushima Fallout May Be Causing Illness in American Babies: Study," *Yahoo News*, April 5, 2013.

Kuznick, Peter. Japan's "Nuclear History in Perspective: Eisenhower and Atoms for War and Peace," *Bulletin of the Atomic Scientists*, April 13, 2011.

Lam, Judy. "Nuclear Plant Operator Rejects Quick Shutdown Demand," *Wall Street Journal*, May 9, 2011.

Lazar, Allan, Dan Karlan, and Jeremy Salter, *The 101 Most Influential People Who Never Lived* (New York: William Morrow Paperbacks, 2006)

"Law to Relieve GE of Liability for Reactors at Fukushima," *Vernon Radiation Safety*, (last visited on Feb. 6, 2014).

Lenssen, Nicholas. *Nuclear Waste: The Problem That Won't Go Away* (Washington, DC: Worldwatch Institute, 1991).

Levey, Stanley. "Nuclear Reactor Urged for Japan T.E. Murray of A.E.C. Tells Steel Union Step is Vital in Atom Race with Russia," *New York Times*, September 24, 1954.

Lewis, Richard. "Even the Best-Laid Plants . . . ," *Citizenvox*, Mar. 24, 2011.

Lochbaum, David et al., *Fukushima: The Story of a Nuclear Disaster*. (New York: New Press, 2014).

Long, Tony. "March 9, 1945: Burning the Heart Out of the Enemy," *Wired*, Mar. 9, 2011.

Lyman, Edwin S. "Letter of Appeal to Samuel L. Bodman Secretary of Energy, U.S. Department of Energy" (Senior Staff Scientist of the Union of Concerned Scientists, July 31, 1997).

Madrak, Susie. "Japan's Response to Fukushima Disaster Was a Mess," *Crooksandliars.com*, Dec. 27, 2011.

Maeda, Risa, and Yoko Kubota, "Japan Releases Radioactive Seawater, Uses Bath Salts," *Reuters/U,S,*, April 4, 2011.

Mariotte, Michael. "On Ninth Anniversary, Ukraine Officials Say 125,000 Have Died From Chernobyl." *Nuclear Monitor*, May 1995.

Maugh, Thomas. "Japan's Prime Minister Criticizes Nuclear Plant's Owner," *Los Angeles Times*, Mar. 29, 2011.

McCurry, Justin. "Fukushima Cleanup Recruits 'Nuclear Gypsies' From Across Japan," *Guardian*, Jul. 13, 2011.

McCurry, Justin. "Nuclear and Industrial Safety Agency Raises Severity Rating to Maximum on International Nuclear and Radiological Event Scale," *Guardian/UK*, April 12, 2011.

McNeill, David. "Japan Must Develop Nuclear Weapons, Warns Tokyo Governor," *The Independent*, March 8, 2011.

"Memorandum from the Director of the Foreign Operations Administration (Stassen) to the Duty Assistant to the President," June 27, 1955, *FRUS*, 1955-1957, Vol. X: p.11.

Mercado, Stephen C. "Book Review: A Limited Japanese view of the CIA," *International Journal of Intelligence and Counterintelligence*, 24: 211-216 (2011)

Meyer, Cordula. "Studying the Fukushima Aftermath: 'People are Suffering from Radiophobia,'" *Spiegel International*, Aug. 19, 2011.

Miller, Jess. "Seaweed Contamination Cause for Concern in Japan," *Greenpeace*, May 12, 2011.

Mirsky, Steve. "Nuclear Experts Explain Worst-Case Scenario at Fukushima Power Plant," *Scientific American*, March 13, 2011.

Mirsky, Steve. "The Japan Earthquake and Tsunami," *Scientific American*, March 12, 2011.

Moore, Joe. *Japanese Workers and the Struggle for Power 1945–1947* (New York: Bell & Howell Information & Lea,1983).

Most, Matthew. "Fukushima: Mark Nuclear Reactor Design Caused GE Scientist to Quit in Protest," *ABC News*, Mar. 15, 2011.

Most, Matthew. "General Electric Defends Nuclear Plant Design," *ABC News*, Mar. 21, 2011.

Mufson, Steven. "Japanese Nuclear Reactors in Peril, Radiation Surges After Earthquake, Tsunami, *Washington Post*, Mar. 12, 2011.

Mufson, Steven, and William Branigin, "Explosion Rocks Japanese Nuclear Power Plant: 5 Reactors in Peril," *Washington Post*, March 11, 2011.

"Murky Past of Japan's Troubled Nuclear Industry Revealed," *The Independent*, Mar. 25, 2011.

Nagasawa, Michio. "Senso to Shimbun [War and Newspapers]," *DIG Newsletter*, August 2007.

Nakasone, Yasuhiro, *Heaven and Earth Have Feelings: Speaking of Fifty Years of Postwar Politics* (Tokyo: Bungei Shunju, 1995).

Nakumura, David. "Japanese Nuclear Plant's Safety Analysts Brushed Off Risk of Tsunami," *Washington Post*, Mar. 24, 2011.

National Academy of Science, *Biological Effects of Ionizing Radiation*, February 19, 1998.

National Archives, "Research Aid: Cryptonyms and Terms in Declassified CIA Files: Nazi War Crimes and Japanese Imperial Government Records Disclosure Acts," June 2007.

Nelson, Craig Doyle. *Nuclear Bonds: Atoms for Peace in the Cold War and in the Non-Western World* (Columbus, Ohio: Ohio State University, 2009).

Nessman, Ravi, and Yuri Kageyama, "Japan Nuclear Crisis: Fukushima Plant Operator Announces Plan To End Crisis," *Associated Press*, April 17, 2011.

NHK: *Gendaishi Sukupu dokyumento: Genpatsu Donyu no Shinario, Reisenka no Tainichhi Genshiryoku Senryaku* [NHK Current Scoop Document: Scenario for the Indoctrination of Atomic Energy—Anti-Japanese Nuclear Strategies during the Cold War], 1994.

"NHK World Edano No Health Risk 20km Away From Plant," *Allvoices*, Mar. 16, 2011.

Nippon TV Network: Shoriki & America's Anti-Communism Foreign and Intelligence Policies," www.f.waseda.jp/tarima/coldwarNTV.pdf.

Nuclear Energy Institute, *Resources & Stats Nuclear Waste: Amount and On-Site Storage*, http://www.nei.org/resourcesandstats/nuclear statistics/nuclear wasteamountsandonsitestorage/ (last visited Feb. 9, 2014).

Nuclear Regulatory Commission, *Reactor Safety Team, Confidential Assessment*, March 26, 2011.

O'Callaghan, Jonny. "Is Nuclear Power Safe?," *How It Works*, March 15, 2011.

Onishi, Norimitsu. "As Japan Works to Patch Itself Up, a Rift Between Generations Opens," *New York Times*, February 12, 2012.

Onishi, Norimitsu, and Martin Fackler. "Anger in Japan Over Withheld Radiation Forecast," *New York Times*, August 9, 2011.

Onishi, Norimitsu, and Martin Fackler. "Utility Reform Eluding Japan After Nuclear Plant Disaster," *New York Times*, November 8, 2011.

Ono, Takuji. *Egakikaerareta Tetsuwan Atomu [Modified Stories of Astro Boy*] (Tokyo: NTT Press, 2008).

Osgood, Kenneth. *Total Cold War: Eisenhower's Secret Propaganda Battle at Home and Abroad* (Kansas: University Press of Kansas, 2008).

"Partial Meltdown in Japanese Nuclear Plant?" *Mystical Politics*, Mar. 13, 2011.

Partner, Simon. *Assembled in Japan: Electrical Goods and the Making of Japanese Consumer* (NY: Columbia University Press, 1999).

Peaslee, S. Claire. "Sending Signals to the Future: How to Design Long-Term Warning Markers for WIPP," *Nuclear Guardianship Forum*, 1993.

Pentland, William. "Risk of Nuclear Catastrophe Escalates in Japan—Worse than Chernobyl," *Forbes*, March 11, 2011.

Peplow, Mark. "Chernobyl's Legacy: Twenty-five Years Alter the Nuclear Disaster, the Clean-up Grinds On and Health Studies Are Faltering. Are There Lessons for Japan?" *Nature*, March 28, 2011.

Phillips, L. "Europe Divided Over Nuclear Power After Fukushima Disaster," *Guardian*, May 25, 2011.

Pilling, David et al. *Tsunami: Japan's Post-Fukushima Future*. (NY: Foreign Policy Magazine, 2011)

Pollack, Andrew. "Japanese Revisit Nuclear Zone While They Can," *New York Times*, April 21, 2011.

"Postwar Defense Aide was CIA Mole," *Japan Times*, October 5, 2009.

"Reactor Core Status of Fukushima Daiichi Nuclear Power Station Unit 1," *TEPCO*, May 15, 2011.

Robson, Seth. "Streets of Tokyo, Yokota AB Remain Quiet Two Weeks After Quake," *Stars & Stripes*, Mar. 25, 2011

"Rokkasho; DeAngelis, Fusako, Rokkasho: Japan's Nuclear Village," *Earth Island Journal*, Spring, 1996.

Ruggiero, Greg and Stuart Sahulka, "Critical Mass: Voices for a Nuclear-Free Future" (presented at the Open Media and The Campaign for Peace and Democracy, Westfield, NJ, 1996).

"Russian Announces Enormous Finds of Radioactive Waste and Nuclear Reactors in Arctic Seas," *Bellona*, Aug. 28, 2012.

Salazar, Adan. "The Good and Bad News About West Coast Radiation," *Alex Jone's Inforwars*, Jan. 14, 2014.

Schwoch, James. "Crypto-Convergence, Media, and the Cold War: The Early Globalization of Television Networks in the 1950s," May 2002, http://cmsw.mit.edu/mit2/Abstracts/MITSchwochTV.pdf.

Sekiguchi, Toko. "Japan, Vietnam to Move Forward on Nuclear Deal," *Wall Street Journal*, October 31, 2011.

"Shinpojumu Fukushima Genpatsu de Naniga Okitaka—Anzen Shinwa no Hokai Hokoku [Symposium: What Truly Happened at Fukushima Nuclear Plant: Deconstruction of Safety Myth]," *Simin Tsushin*, 2012, http://archives.shiminkagaku.org/archives/csijnewsletter_015_nagasoe_121225.pdf.

"Shusenno Hini Gensiryoku no Heiwariyo o Kangaeru [To Consider "the Atoms for Peace" Program at the Anniversary of the End of WWII]," *Tanemaki Journal*, April 15, 2011.

Sixth Report of Royal Commission on Environmental Pollution: Nuclear Power and the Environment (London, England: Her Majesty's Stationery Office, 1976).

Smith, Jim, and Nicholas A. Beresford, *Chernobyl: Catastrophe and Consequences*. Berlin, Germany: Springer/Verlag, 2005.

Smith, J. T. et al., "Pollution: Chernobyl's Legacy in Food and Water," *Nature*, May 11, 2000.

"Soaring Radiation Spikes in Water At and Around Fukushima Daiichi Hindering Progress to Cool Reactors," *Bellona*, Mar. 28, 2011.

"SPEEDI Yosoku 'Kohyo Dekinai" Monbusho Bunsho ni Kisai [SPEEDI Prediction Impossible to Disclose: Recorded in MEXT Documents," *Kyodo*, Mar. 3, 2012.

Sternglass, Ernest J. *Low-Level Radiation* (London: Earth Island Limited, 1973), pp.6–9.

Sternglass, Ernest J. "Leukemia and Ionizing Radiation," in Low-Level Radiation (London: Earth Island Limited, 1973).

Sugita, Hiroki. Postwar Propaganda Campaign: USIS Role Revealed in Japan's Tilt Toward West, *Japan Times*, November 21, 1997.

"Summary Report of RSMC Beijing on Fukushima Nuclear Accident Emergency Response," *World Meteorological Organization*, Oct. 31, 2011.

"Survivors Commemorate Asian Tsunami," *IntellAsia*, Dec. 27, 2012.

Tabuchi, Hiroko. "In Japan, a Tenuous Vow to Cut," *New York Times*, September 1, 2011.

Tabuchi, Hiroko. "Japan Acts Alone to Weaken Its Currency," *New York Times*, October 31, 2011.

Tabuchi, Hiroko. "Japan Lifts Ban on Beef from Disaster Area," *New York Times*, August 25, 2011.

Tabuchi, Hiroko. "Japan Starts Nuclear Cleanup, With Little Idea of How," *New York Times*, February 10, 2012.

Tabuchi, Hiroko. "Japan's Premier Says He'll Keep Nuclear Phase-Out," *New York Times*, September 3, 2011.

Tabuchi, Hiroko. "Japanese Island's Activists Resist Nuclear Industry's Allure," *New York Times*, August 28, 2011.

Tabuchi, Hiroko. "Japanese Parliament Backs Noda as Prime Minister," *New York Times*, August 30, 2011."Environmental Radiation: Where the Radiation Went," *Nature*, September 1, 2011.

Tabuchi, Hiroko. "Japanese Tests Find Radiation in Infant Food," *New York Times*, December 6, 2011.

Tabuchi, Hiroko. "Mounting Woes Prompt Sony to Predict Another Annual Loss," *New York Times*, November 2, 2011.

Tabuchi, Hiroko. "Radiation Poses Barrier to Repair Work at Plant," *New York Times*, April 18, 2011.

Tabuchi, Hiroko. "Strong Yen Is a Two-Edged Sword for Japan," *New York Times*, August 19, 2011.

Tabuchi, Hiroko. "With Reactor Damage Thought to Be Worse, Tokyo Utility Sticks to Plan," *New York Times*, May 17, 2011.

Tabuchi, Hiroko. "World/Asia Pacific: Fukushima Meltdown in Japan May Have Been Worse Than Thought," *New York Times*, November 30, 2011.

Tabuchi, Hiroko. "World/Asia Pacific: Japan Joining Talks on Trans-Pacific Partnership, Noda Says," *New York Times*, November 12, 2011.

Tabuchi, Hiroko & Ken Belson, "Contaminated Water Escaping Nuclear Plant, Japanese Regulator Warns," *New York Times*, Mar. 28, 2011.

Tabuchi, Hiroko and Martin Fackler. "More Radioactive Water Leaks at Japanese Plant," *New York Times*, December 4, 2011.

Tabuchi, Hiroko, and Matthew L. Wald, "Japanese Scramble to Avert Meltdowns as Nuclear Crisis Deepens After Quake," *New York Times*, March 12, 2011.

Takahara, Kanako, and Kazuaki Nagata, "High Radiation Found Outside No-go Zone; but Edano Says Evacuation Area Won't Be Expanded for Time Being," *Japan Times*, April 1, 2011.

Taketani, Mitsuo. *Taketaniya Mituo Gendaironshu* [Collection of Modern Writings by Mitsuo Taketani] (Tokyo: Keiso Shobo, 1974).

Talmadge, Eric, and Mari Yamaguchi, "Death Toll Tops 10,000, Multiple Nuclear Meltdowns Loom," *Associated Press*, Mar. 13, 2011.

Tanikawa, Miki. "Finding Use for 25 Million Tons of Rubble," *New York Times*, June 2, 2011. "Can It Happen Here? Yes—if We Let It. Fukushima's Lessons for the United States," *Catalyst*, Summer 2011.

Tanaka Yuki, and Peter Kuznick, "Japan, the Atomic Bomb, and the 'Peaceful Uses of Nuclear Power," *The Asia-Pacific Journal*, vol. 9, issue 18, 1, May 2, 2011.

Taylor, Phil. "Glossary of relevant Terms and Acronyms," The Institute of Communications Studies, University of Leeds, 1987.

Taylor, Theodore B. "Nuclear Weapons, Nuclear Power and Solar Energy," in Gred Ruggiero and Stuart Sahulka, *Critical Mass: Voices for a Nuclear-Free Future*, Wesfield, NJ: Open Media and the Campaign for Peace and Democracy, 1996).

"TEPCO Confirms Extra Partial Fuel Rod Meltdown at Plant," *News Business* (BBC, May 23, 2011).

"TEPCO May Face $23.6bn in Claims, JP Morgan Says," BBC/Business/News, April 12, 2011.

"TEPCO to Invest in South Texas Project Expansion (STP 3&4): First Japanese Utility to Invest in the Overseas Nuclear Power Project, *TEPCO*, May 10, 2010, http://www.tepco.co.jp/en/press/corp-com/release/10051001-e.html.

"TEPCO Says Plutonium Found on Quake-Damaged Plant Grounds," CNN World, Mar. 28, 2011.

"The Department of Defense Role in Foreign Assistance Background, Major Issues, and Options for Congress," *Congressional Research Service*, December 9, 2008.

"Timeline: Japan Power Plant Crisis," BBC News, Mar. 13, 2011.

"Tokyo, Japan," Associated Press, March 20, 2011.

Tomooki Yasuda, "A Report Suggests a Second Meltdown at Reactor at Fukushima Plant," *Ashia Shimbun*, August 22, 2011.

"Typhoon Rains Kill at Least 25 and Maroon Thousands in Japan," Associated Press, September 4, 2011.

"Ultimate Bunker Buster: U.S. Military Speed Plans for 13-Ton Bomb," *Daily Mail*, Oct. 13, 2009.

"U.S. Concerned Japan Facing Situation That Could Be Deadly for Decades," *ABC News*, March 16, 2011.

U.S. Department of *Energy, Dep't* of Energy to Send Proposed Yucca Mountain Legislation to Congress (Apr. 4, 2006).

"U.S. Embassy Stops Distributing Iodide Pills to Citizens in Japan," *Stars & Stripes*, Apr. 18, 2011

U.S. Environmental Protection Agency, "Cesium," *EPA/Radiation Protection*, March 6, 2012, http://www.epa.gov/radiation/radionuclides/cesium.html.

U.S. Environmental Protection Agency, "Iodine," EPA/Iodine, March 6, 2012, http://www.epa.gov/radiation/radionuclides/iodine.html.

U.S. Environmental Protection Agency, "Yucca Fact Sheet #1," http://www.epa.gov/radiation/docs/yucca/epafactsheet1v3.pdf (last visited Feb. 9, 2014);

"U.S. Forces Given SPEEDI Data Early," *Japan Times*, Jan. 18, 2012.

"U.S. Sailors Sue Japanese Nuclear Plant Owner TEPCO," *Environmental News Service*, Jan. 3, 2014.

van Wolferen, Karel. *The Enigma of Japanese Power* (New York: Knopf, 1990)

Wagner, David. "Report From Tokyo: Who to Believe?" *Huffington Post*, May 27, 2011.

Wald, Matthew. "NRG Abandons Project for 2 Reactors in Texas," *New York Times*, April 19, 2011.

Walt Disney. *Walt Disney's Wonderful World of Color: Our Friend the Atom* (NY, Walt Disney, 2011).

Wasserman, Harvey. "Humankind's Most Dangerous Moment: Fukushima Fuel Pool at Unit 4: "This is an Issue of Human Survival," *Global Research*, September 20, 2013.

Wassener, Bettina. "Moody's Warns That It Might Downgrade Japan's Debt," *New York Times*, May 31, 2011.

Weiss, Thomas. "Dangers and Health Effects of Depleted Uranium," *Disabled World*, Sep. 4, 2013.

West, James. "Uranium Stocks Under Siege as Fukushima Continues Spewing Radiation," *Midas Letter*, May 30, 2011.

"White Smoke Around the Fukushima Daiichi Nuclear Power Station Unit 3," *TEPCO*, Mar. 14, 2011.

"Will US Dump Nuclear Waste at Sea Again?" *Christian Science Monitor*, Apr. 8, 1982.

Wing, Steve et al., "Mortality Among Workers at Oak Ridge National Laboratory," *Journal of the American Medical Association* Vol. 265, no. 1(March 20, 1991): 1397–1402.

"Without SPEEDI, Residents Fled in Direction of Radiation," *Asahi Shimbun*, Dec. 27, 2011.

Yabloskv, Alexey, Vassily Nesterenko, and Alexey Nesterenko, *Chernobyl: Consequences of the Catastrophe for People and the Environment* (New York: The New York Academy of Sciences, 2009).

Yamaguchi, Mari. "Japan Nuclear Reactor Damage Worse Than Expected," *Huffington Post*, May 12, 2011.

Yamaguchi, Mari. "Japan Population to Shrink by One-third by 2060," MSN News/Associated Press, January 30, 2012.

Yamaguchi, Mari. "Radioactivity Rises in Sea Off Japan Nuclear Plant," Associated Press, April 16, 2011.

Index

About the Authors

Richard Krooth, a legal scholar and political economist, has practiced law and taught sociology at the University of California, Riverside. He is presently a research associate at the University of California, Santa Cruz.

Morris Edelson was editor of *Quixote* during the years of the counter-culture and anti-war movement at the University of Wisconsin in Madison and has taught at Houston Community College.

Hiroshi Fukurai is professor of sociology and legal studies at the University of California, Santa Cruz.

CPSIA information can be obtained at www.ICGtesting.com
Printed in the USA
BVOW03*1944010215

385636BV00004B/5/P